MATHEMATICS
A CRASH COURSE

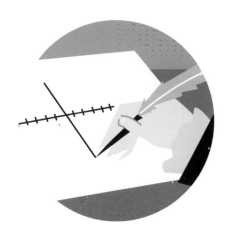

MATHEMATICS
A CRASH COURSE

BECOME AN INSTANT EXPERT!

Brian Clegg &
Dr. Peet Morris

METRO BOOKS
New York

METRO BOOKS
New York

An Imprint of Sterling Publishing
1166 Avenue of the Americas
New York, NY 10036

ISBN: 978-1-4351-6966-1

For information about custom editions, special sales,
and premium and corporate purchases, please contact
Sterling Special Sales at 800-805-5489
or specialsales@sterlingpublishing.com.

Manufactured in China

2 4 6 8 10 9 7 5 3 1

sterlingpublishing.com

Credits: Art Director James Lawrence;
Design JC Lanaway;
Illustration Beady Eyes;
Series Concept Design Michael Whitehead

INTRODUCTION

We don't know how the basics of math were first devised, as it predates written language and history. The simple ability to be aware that there is a difference between, say, one object and two is likely to have been innate in human ancestors long before *Homo sapiens* emerged around 200,000 years ago. We can say this because it has been shown that dogs have this kind of awareness. When a trick is performed on a dog so that it thinks that two items of food have been put in its bowl, but in reality there is only one, it does the canine equivalent of a double take.

Such an awareness of numbers without any concept of counting, then, seems to have been with us for a long time. However, it is likely to have been when humans began to have assets, to trade, and to build, that a more concrete form of math became necessary. The specific trigger was likely the formation of early cities and the development of writing—without notation, math could only ever be trivial.

The earliest mathematics

The oldest example we have of math being written down is 6,000 years ago, in the city-state of Uruk, in present-day Iraq. Uruk was at the center of the Sumerian civilization, and with unprecedented numbers of people living together, the need to keep track of foodstuffs and other tradable goods was key. And as buildings became more sophisticated, measurements were needed for their construction. However, the people of Uruk were yet to equate abstract numbers with collections of objects. Today, we can deal with a concept like "four" as a number and apply it equally to anything from the Four Horsemen of the Apocalypse to the four legs on a dog. But in Uruk, the inhabitants considered some objects to be so different that they needed alternative representation. So, while they used the same numbers to count live animals and dried fish, they used a different set for cheese, grain, and fresh fish.

As civilizations became more established, new concepts arose that were increasingly removed from reality, notably in Greece, India, and China. Later, this sophisticated form of math would move via the Arabic-speaking countries to Europe, and eventually the US. Only with the step back from pragmatic application to thinking about what was actually happening in the math itself was it possible for math in the modern sense to be born. The Sumerians knew, for example, that a right (or right-angled) triangle—so useful when surveying a piece of land—had a ratio of sides such that if one of the shorter sides was three units and another four, then the longest side would be five units. They knew this from observation, but proofs that this was exactly the case for all right triangles came much later.

Building the mathematical toolkit

For many uses, the math developed over 2,000 years ago largely serves today. No one needs topology to deal with the basics of everyday life. But from the sixteenth century onward, the possibilities for applying math blossomed, becoming applicable not just to reality, but to what might be—"probability," the mathematics of chance, for instance, would transform our ability to think about the future.

Early mathematicians had dealt only with futures where there were distinct rules. Given, say, a known rate of interest, it was possible to calculate exactly how much money you might have in a bank account in a year's time (if nothing was added or removed). These were very artificial conditions. The real world was full of unknowns. But, as US defense secretary Donald Rumsfeld would famously observe, there are both known unknowns and unknown unknowns. Probability would prove an ideal mechanism for dealing with known unknowns. We don't know what will come up when we toss a coin, but we do know that, with a fair coin, there's a 50 percent chance of a head and a 50 percent chance of a tail.

Sometimes a new field of math opens up when there is a clear application. In the seventeenth century, Isaac Newton and Gottfried Leibniz devised the techniques now collectively known as "calculus" in order to deal with specific problems—how to work out, for example, the distance traveled when a moving body is accelerating. However, in many cases, mathematicians were simply happy exploring their mathematical universe without any thought of an application. Yet surprisingly often, such apparently pointless work would later have major practical applications.

All manner of mathematical oddities have eventually proved useful. "Imaginary numbers"—the square root of negative numbers—were devised long before applying them; the sixteenth-century Italian mathematician Girolamo Cardano commented that an imaginary number was "as subtle as it is useless." Yet by the nineteenth century they were widely employed by both physicists and down-to-earth electrical engineers, and continue to be used today.

Another example is the importance of math in the history of computing. Originally, computers were people. Up to the 1940s, the term largely applied to individuals who performed manual calculations with pencil and paper. The potential advantages of mechanizing the work of a computer were clear for centuries, and there were early attempts with mechanical calculators, but electronic computers—first developed in the 1940s—took the mathematical content to a whole new level.

Math and the digital era

Three huge mathematical leaps were required to develop the electronic infrastructure at the heart of the modern world. The first was using binary digits: no longer working with everything from 0 to 9, but simply with 0s and 1s. Humans could, of course, have always done their mathematics this way—but it would have been tedious in the extreme. However, for a computer there is no disadvantage, and a huge positive. The values 0 and 1 can be represented very easily by physical things. A switch that is on or off. A device that either has an electrical charge or no charge. Either/or situations are easily represented by electronic components.

Secondly, computers are designed around mathematical logic. Those same 0 or 1 values can also be considered as false or true. And in the nineteenth century, a mathematical approach to logic had already been developed. This would underpin the architecture of the innards of computers: they are, in essence, huge logic engines, connecting together vast numbers of very simple logic "gates"—mechanisms for making logical combinations or transformations of those 0s and 1s.

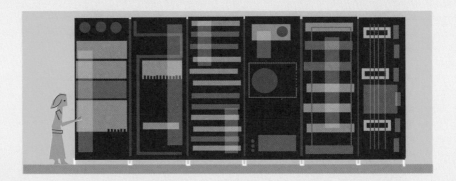

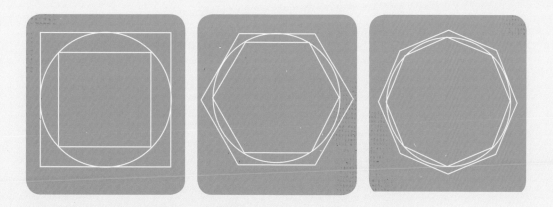

The final mathematical component came when electronics moved from vacuum tubes to solid-state devices such as transistors. The design of such devices required an understanding of quantum mechanics, the physics of the very small particles that, for example, make up matter. And quantum mechanics, like most modern physics, is extremely mathematical. Not only did solid-state design require an understanding of the tools of higher mathematics, quantum physics has probability at its heart. Without a profound understanding of mathematics, the modern computer could never have been constructed.

The unreasonable effectiveness of math

It's a sad truth that math has the reputation of being "difficult." Part of the problem is that many of us simply don't speak the language. To a mathematician, an equation is a compact, efficient way to put across a relationship that would be far less comprehensible in words. But to many of us, the merest sign of x's, y's, and other symbols is an impenetrable mess that our eyes bounce off.

There is a story that the physicist Eugene Wigner told in a paper exploring the "unreasonable effectiveness of mathematics in the natural sciences." Wigner describes two school friends, meeting for the first time after leaving school. One works in mathematics and shows his friend a paper he has written on population change. The mathematician tries to explain what all his strange symbols mean in this document containing very few words.

His friend is doubtful. How can such a bunch of squiggles bear any resemblance to what actual people do in the real world? The mathematician, trying to reassure his friend, points to a π symbol in the equations. "That's pi. You know what that is? The ratio of the circumference of a circle to its diameter." The friend shakes his head. "Now I know you're messing with me. What has the population got to do with the circumference of a circle?"

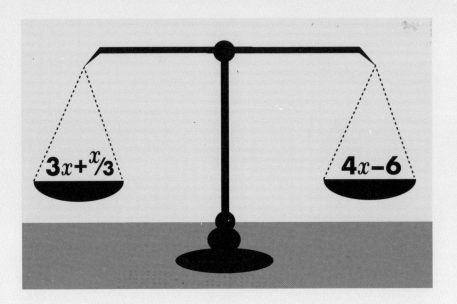

Math is, in the end, an abstract discipline operating in its own world. Although it may have started with a degree of one-to-one correspondence between, say, numbers and physical objects, it soon set off on its own path with the introduction of concepts like negative numbers, which it's harder to envisage as real objects, or irrational numbers, such as the square root of 2. An irrational number is one that can't be represented by a ratio of any two whole "real" numbers, making it much harder to link directly to reality in the mind.

In math, anything goes, as long as it is consistent. Mathematicians can decide that $1 + 1 = 3$ or that there are 27,319 dimensions of space. However, as Wigner points out, despite its arbitrary nature, mathematics has an apparently unreasonable ability to be supremely useful in a real world where we can't make up our own rules.

It would be impossible to overstress how important mathematics is to the modern, technology-driven world. Of course, that's obviously the case when we buy something or do our banking, but it's far more. Physics, the science behind most technology, became utterly dependent on mathematics in the nineteenth century, and is now often nearly indistinguishable from pure mathematics. And, as we have seen, the information and computing technology at the heart of today's modern economy and life is entirely driven by mathematics. We see the physical computer or cell phone—but inside the digital technology it is the manipulation of numbers and logic that makes everything happen. Living in a digital world inherently means living in a mathematical world.

The biggest problem we have with math is the way that it's taught in school. Once we've learned the basics that we'll need for the everyday handling of money and measurements, we go on to study the kind of math that will be useful if we subsequently work in the sciences or engineering. But for the vast majority, this will never be the case—giving the impression that higher math is a waste of time. If we were provided with an engaging overview of what math is and what it can do, without having to solve simultaneous equations or prove geometric theorems, then far more of us might get the point of it. That's exactly what this book is designed to do.

This book is divided into four chapters, each covering a major developmental route in the topic, and containing 13 short articles. That's an article a week for a year—but feel free to consume them at your own speed. However you chose to look at it, though, this book takes a subject that can seem intimidating and makes it approachable. So let's get started with Chapter number 1 . . .

How to use this book

This book distills the world of mathematics into 52 manageable chunks, allowing you to skim-read or delve in a bit deeper. There are four chapters, each containing 13 topics, prefaced by biographies of key figures, alongside a timeline of significant milestones. The introduction to each chapter gives an overview of the key concepts you might need to navigate.

Each topic has three paragraphs.

The Main Concept provides a theory overview.

TRIGONOMETRY

THE MAIN CONCEPT | The relationship between angles and the length of sides of triangles is a natural extension of geometry, called "trigonometry," meaning "triangle measurement." Thinking of one of the angles in a right triangle, the "sine" of that angle is the ratio of the length of the opposite side to that of the hypotenuse (longest side). The "cosine" is the same measure but using the ratio of the third side to the hypotenuse. And the "tangent" is the ratio of the opposite side to the third side (you may be familiar with the "SOH CAH TOA" method of remembering this). These simple "functions" have proved useful in surveying, navigation (for example, with sextants), and astronomy. Trigonometry also enables triangulation—deducing distances by measuring the angles of triangle between known positions and a new point, which proved transformational for accurate map making. However, the triangular definition of the trigonometric functions was limiting as it could only deal with angles up to 90 degrees. An extended definition imagines the triangle drawn in a circle with the right angle at the center. The hypotenuse is the radius and can be swept around the circle, giving values that oscillate periodically. This approach produces an alternative unit of angle, the "radian." Trigonometry was also extended out of the plane, using a sphere rather than a circle as the basis.

DRILL DOWN | Once trigonometric functions were measured using a circle, it became possible to think of the size of an angle as the amount by which the hypotenuse of the triangle has swept around the circumference of a circle, based on a circle with a radius of 1; the circumference is 2π—so a rotation of 360 degrees is considered 2π radians; the same goes to this new unit. This approach proved profitable when trigonometry was combined with calculus, as using the radius measure produced simpler results. It is also a more mathematically rigorous unit than the culturally influenced degrees. Radians are now the standard scientific unit for the measurement of angles.

FOCUS | In many countries, high points in the landscape are marked with different styles of concrete pillars, usually signed with a scary plaque. Known as "triangulation stations" in the US and "trig points" in the UK, these are mounts for theodolites, optical instruments used to enable trigonometric measurements. They have largely been made redundant by GPS satellite navigation systems.

The Drill Down functions as a critique of the main concept, or looks at one element of the main concept in more detail, to give another angle or enhance understanding.

The Focus provides a counterargument or an alternative viewpoint from a key player in the field, or a key event subsequent to the initial theory.

"**Mathematics attempts to establish order and simplicity in human thought.**"

EDWARD TELLER,
THE PURSUIT OF SIMPLICITY (1981)

1
ARITHMETIC
& NUMBERS

INTRODUCTION

Counting is likely to have begun with fingers and thumbs. This accounts for both the familiar decimal approach, with one number per digit (hence the use of the term for numbers as well as parts of the anatomy) and the use of "base 60" by the Sumerian and Babylonian civilizations. Where decimal (base 10) numbers run from 0 to 9 before starting a new column, base 60 numbers run from 0 to 59 before restarting. Counting to 60 on two hands may seem a stretch, but it has been suggested that this was done by touching a thumb to each of the 12 finger joints on one hand in turn, while using the other hand to count off 5 lots of 12. (Of course, the same approach could have been used for base 144, counting 12 on each hand, but thankfully no one seems to have thought of this.)

The birth of numbers

Soon, hands would have been supplemented by markers—tally marks carved on bones or wood, or groups of stones—which would eventually transmute into that powerful early calculating device, the abacus. But the step that took us from counting to arithmetic was the introduction of numbers—symbols and words representing these finger positions or tally mark combinations. Initially this may have just been a convenience—it's easier to say "five" than to show someone a hand with the fingers all in the right position—but with the advent of writing, numbers became essential.

Once numbers were regarded as things in their own right, it became clear that they fell into different categories. There were even numbers, which could be split into two equal amounts, and odd numbers that couldn't be so divided. As we will see in Chapter 2, early mathematicians were very interested in shapes because of the importance of being able to divide up land and design buildings. It was also noticed that some numbers seemed to correspond to shapes. So, for example, 6 was a triangular number: envisage a set of pebbles arranged in rows of 1, 2, and 3 beneath each other; it can easily be made into a triangle. Other numbers made squares: 4, 9, and 16, for example.

Working with numbers

With numbers given names and symbols it also became easier to consider mechanisms for manipulating those numbers—combining them and interacting with them, just as it was possible to manipulate physical objects. Sometimes discoveries would be made that had no practical application but were nonetheless entertaining. The members of the Pythagorean School of ancient Greece, for example, were aware of linked pairs of numbers known as "amicable" numbers.

Understanding amicable numbers required another arithmetical concept—the factor. The factors of a number are the values it can be divided by and still leave a whole number. So, for instance, the significant factors of 12 are 1, 2, 3, 4, and 6; 5 isn't a factor because $^{12}\!/_5$ isn't a whole number, and anything bigger than 6 can't be a factor as it would produce a result smaller than 2. (Strictly, the number itself is a factor, but this is considered trivial.) Amicable numbers are those where the factors of one number add up to the other, and vice versa. The best known are 220 and 284: add the factors of 220 and you get 284, and those of 284 give 220.

Once the concept of factors was understood, another strange set of numbers emerged—the primes. These are numbers that only have 1 and themselves as factors. Primes would repeatedly prove valuable in the development of higher mathematics, yet even today the way they are distributed cannot be predicted. With time, other interesting types of numbers were discovered, too. Division implied ratios, but there also seemed to be non-whole numbers that weren't ratios, or fractions—the so-called irrational numbers, of which pi is the best-known example. With each extra feature, the power of numbers seemed to increase. What had once been little more than a way of checking nothing had been lost or stolen was transformed into a parallel world of numbers and their manipulation, which could be sent on wild journeys of its own. Mathematics was taking off.

TIMELINE

SUMERIAN NUMBERS
The Sumerian numeral system, later taken over and developed by the Babylonians, is introduced. It is base 60, rather than the now-familiar base 10—and uses notations positioned in columns to indicate multiples of 60. We still use base 60 to measure time in for minutes and seconds.

BINARY NUMBERS
The Indian mathematician Pingala makes the first known reference to binary, or base 2, numbers. He also refers to Pascal's triangle (a triangular arrangement of numbers starting with three 1s, then producing numbers below by adding adjacent numbers above) and the Fibonacci series, all to explain the meter of Sanskrit writings.

At least 35,000 BCE

c. 3000 BCE

c. 530 BCE

c. 200 BCE

TALLY STICKS
The Lebombo bone—a carved bone from a baboon's leg—is the oldest near-certain tally stick, featuring 29 notches. It is found in the Lembombo Mountains bordering South Africa and Swaziland. Some date it to even further back—around 41,000 BCE.

THE PYTHAGOREANS
The School of Pythagoras is founded in Croton, southern Italy. With an obsession with whole numbers, including a belief that the whole universe is based on whole numbers and ratios, the Pythagoreans take numbers from a simple tally mechanism to a separate concept that can be operated in its own right.

NEW NUMBER SYSTEMS

Fibonacci (a nickname meaning "son of Bonacci") writes *Liber Abaci* ("Book of Calculation"), which introduces both zero and Indian/Arabic numbers to Europe. The new numbers are resisted at first, in part because accounts can be fiddled by changing 0 to 6 or 9.

LOGARITHMS

John Napier publishes *Mirifici Logarithmorum Canonis Descriptio* ("Description of the Wonderful Rule of Logarithms"), which makes use of logarithms, simplifying multiplication and division by using the exponents that can be added or subtracted. They become the standard vehicle for calculation before mechanical and electronic devices take over.

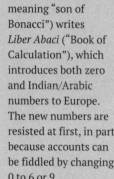

628 CE	1202	1545	1614

TRUE ZERO

The Indian mathematician Brahmagupta makes the first known use of "true" zero (the value of a number taken away from itself). Zeroes used as placeholders to keep numbers in columns in a positional system date back much earlier to the Babylonians, who started to use \\ for an empty space around 1,200 years before Brahmagupta's work.

NEGATIVE NUMBERS

Publication of Girolamo Cardano's book *Ars Magna* ("The Great Art"). Cardano is the first to explicitly manipulate negative numbers and to consider (if partially and inaccurately) the implications of having the square root of a negative number, now known as an imaginary number.

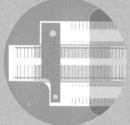

BIOGRAPHIES

PYTHAGORAS (c. 570–c. 495 BCE)

For most of us, the ancient Greek philosopher Pythagoras is only associated with the geometrical theorem about the square on the hypotenuse. In fact, though, this predates him. It's difficult to pin down biographical details, or to separate his work from the wider school he founded, but Pythagoras was born on the island of Samos, and around the age of 40 founded a school at Croton in the south of Italy. The Pythagorean School believed that the universe was driven by whole numbers; the school is said to have had the words "All is number" carved on its door lintel. Where basic use of numbers in counting probably originated in everyday transactions, intimately linked to physical objects, the Pythagoreans made numbers an abstract ideal. Arguably it was they who broke mathematics away from its direct link to reality, making it more independent. Whole numbers were given particular attributes, and patterns of numbers were studied in everything from the stars to musical notes. According to legend, when one of the followers of Pythagoras, Hippasus, threatened to reveal to the world that the square root of 2 could not be made with any ratio of whole numbers, he was drowned.

GIROLAMO CARDANO (1501–1576)

The Italian mathematician Girolamo (or Gerolamo) Cardano, often known by the French version of his name, Jérôme Cardan, was born in Pavia. He studied medicine at the University of Pavia and went on to practice (without a license) in the town of Saccolongo. From there he moved to Milan, where he continued with medicine but also taught mathematics. Cardano stretched the boundaries of the concept of numbers, working with negative numbers and even touching on imaginary numbers. He was also one of the first to seriously study the basics of probability, though his book on the subject, *Liber de Ludo Aleae* (*Book of Games of Chance*), was not published until long after his death. It seems that anything so closely associated with gambling was considered unsuitable for publication. Despite being dated, the book helped kickstart the study of probability. Cardano later moved to Bologna, where he was arrested by the Inquisition in 1570. Several months later he was freed and moved to Rome, where, in contrast, the Pope gave him a lifetime annuity. He remained in Rome for the rest of his life, continuing to practice medicine and writing on philosophy.

CARL FRIEDRICH GAUSS (1777–1855)

Though not a familiar name outside math and physics, the German mathematician Carl Friedrich Gauss was one of the greatest pre-nineteenth-century mathematicians. Born in Brunswick to a poor family, Gauss showed mathematical skills from an early age. According to legend, aged 8 he amazed his school teacher by adding up every number from 1 to 100 in seconds (he realized that the series was just 50 pairs of 101). By the time Gauss was at university—sponsored by the Duke of Brunswick, who had been made aware of Gauss's mathematical ability—he was already breaking new ground. He toyed with a career in philology, but the appeal of making new proofs and discoveries won him over to math. Gauss's contributions to mathematics fit into every chapter of this book. For this chapter he contributed to number theory, and particularly modular arithmetic. He was one of the first to think about non-Euclidean geometry and topology. He proved what became known as the "fundamental theorem of algebra," and he developed applications of mathematics in astronomy, optics, and the study of magnetism.

GEORG FRIEDRICH BERNHARD RIEMANN (1826–1866)

Although German mathematician Georg Friedrich Bernhard Riemann made a wide range of contributions to mathematics, he is best remembered for something he failed to prove. Riemann studied theology at the University of Göttingen, but there his ability in mathematics blossomed, and he transferred to the mathematically oriented University of Berlin. One of his earliest developments was Riemannian geometry, which applied calculus to smoothly varying multidimensional surfaces (Einstein had great trouble understanding this sophisticated mathematics in relation to the curvature of space-time when developing his masterpiece, the general theory of relativity). Riemann contributed to the detail of calculus and Fourier series. But it was work on the prime counting function, approximating the number of prime numbers in a range, that led to the Riemann hypothesis. This is a complex mathematical conjecture related to the distribution of prime numbers that, were it true, automatically generates a whole range of proofs in number theory. One of the most significant unsolved mathematical problems, a $1 million prize awaits its solution.

TALLIES

THE MAIN CONCEPT | The tally is almost certainly the oldest known application of mathematics, with the purpose of keeping track of objects, or things that have happened. For a tally, a mark is made—for example, on a piece of wood, stone, or bone—corresponding to each of the items or events being monitored. The very earliest precursors of tallies probably involved using either fingers, or simple objects that could be used as markers, such as sticks or pebbles. However, the problem with these was that they were ephemeral. You only have to make use of your hands for something else, or move the tally objects, and the record is lost. By carving the tally into a durable substance, a tally could be kept indefinitely; later they would frequently take the form of marks made with a stylus in clay, which could be baked hard. A tally might be used to keep score in a game, or to record the number of livestock or provisions in someone's possession. Tallies are also invaluable for recording trades—marks can be added to or removed from a tally when items are loaned to someone else or traded for other goods and services. All of these activities can be carried without ever introducing the concept of numbers.

DRILL DOWN | Though it is unlikely to have occurred to to its early users, the concept of a tally can be expressed in terms of what modern mathematicians call set theory (see page 106). Two sets of items (just collections of things) are the same size if the items can be put in one-to-one correspondence, pairing off an item from one set with an item from the other until all are used up. So, for example, imagine a set of major compass directions and a set of seasons. You could pair off north with spring, east with summer, south with fall, and west with winter. In each case, all the items would be used up, so you would know the sets were the same size without ever knowing how big they were.

FOCUS | *Although the short-lived tallies of fingers or pebbles are long gone, some very ancient markings have been found that are almost certainly tallies. The Lebombo bone (see page 16) is the oldest known tally. A clearer example, with more intricate groupings of notches, is the Ishango bone, from around 20,000 BCE.*

NUMBERS

THE MAIN CONCEPT | Numbers started as a useful shorthand. In verbal communication, having a word for, say, "five" is much easier than saying "Here's a bag of corn, and another, and another, and another, and another." Once written records were used, numbers also made bookkeeping simpler, capable of keeping track of a collection of values that changed with time. The method of representing numbers emerged from tally marks (see page 20), and many numeral systems still use I to represent one, corresponding to a single mark. However, tally marks quickly become unwieldly, so extra symbols were introduced across the world. The crudest approach was the Roman system. Reflecting the fingers, it introduced just two symbols between 1 and 10—V for 5 (one hand), and X for 10 (two hands). Like a tally, there was no significance to the position of the symbols. And yet the Romans did not lack precedents in this respect. The Sumerians and Babylonians had used a sophisticated system where the position of the symbol indicated the size of the number (see page 24), just as we do today. Similarly, Chinese and Indian cultures used positional systems, with the Indian form, modified in translation via Arabic, forming the basis for our present system.

DRILL DOWN | In formal mathematics, numbers correspond to sets of items. The number 0 is an empty set. The number 1 is a set containing the empty set—it has one thing in it (the empty set); 2 is the set containing the previous set . . . and so on. We can then can match real objects with members of the set to assign numbers to them. So, if I have some apples and can pair one of them with the first member of the number 2 set and one with the next member of the number 2 set and there are no more apples or members of the set, there are two apples.

FOCUS | *"Arabic" numerals came to Europe in the thirteenth century, in a book by Italian mathematician Fibonacci, but they could have arrived sooner. Bishop Severus of Syria noted in 662 BCE that Indians had remarkable abilities of calculation and excelled at astronomy. He noted that their approach "surpasses description," and that "this computation is done by means of nine signs" (before the widespread use of zero).*

Brahmi		—	=	≡	+	N	ℓ	٦	٢	?
Hindu	०	१	२	३	४	५	६	७	८	९
Arabic	٠	١	٢	٣	٤	٥	٦	٧	٨	٩
Medieval	O	I	2	3	8	৭	6	٨	8	9
Modern	0	1	2	3	4	5	6	7	8	9

NUMBER BASES

THE MAIN CONCEPT | We use a positional system of numbers in base 10 (where the column a number is placed in is significant). When we write out a number such as 6923, the position of each number indicates how many 10s it should be multiplied by. In this case, the 3 is multiplied by 1 (or 10^0), the 2 by 10 (10^1), the 9 by 100 (10^2), and the 6 by 1000 (10^3). Base 10 is a natural choice as it reflects the digits on our hands, but it is not the only option. Ancient Sumerians and Babylonians used base 60 (shown opposite). In this system the rightmost column was for numbers up to 59, the next left for multiples of 60, the next left for multiples of 3600, and so on. We now only use base 60 for seconds and minutes. We also make wide use of binary, or base 2, numbers. These are far easier for computers to handle, both logically and physically with electronic circuits, which can be either electrically charged or not (written 1 or 0). Logical values of true or false can be handled by treating, for example, 1 as true and 0 as false. Binary is often translated to other bases for human consumption—programmers often use base 16 (hexadecimal) and, less frequently now, base 8 (octal), as explained opposite.

DRILL DOWN | Binary numbers, used by computers, rapidly become unwieldy (it takes a while to realize that 11111100010 is 2018), so programmers translate binary to octal or hexadecimal. The now-outmoded octal system takes bits in groups of 3 and replaces them with 0 to 7. So 11111100010 or 11-111-100-010 becomes 3742. Hexadecimal is trickier as it requires 16 digits, but we only have 10 numerals, so the letters A to F are added, where A is 10, B 11, and so on. For this, a binary number is divided into blocks of four bits: 11111100010 or 111-1110-0010 is 7E2. Hexadecimal is common as modern computers split data into chunks of 16, 32, or 64 bits—easier to divide into 4s than 3s.

FOCUS | *Most number systems have positive integer (whole number) bases, but this is not essential. For example, with a base of −10, there is no need for the "−" sign. As a negative multiplied by a negative is a positive, as with $(-10^2) = +100$, and where $(-10^3) = -1000$, alternate columns indicate positive and negative values. In base −10, 174 is the equivalent of $1 \times 100 - 7 \times 10 + 4 = 34$.*

BASIC ARITHMETIC: NEGATIVE NUMBERS

THE MAIN CONCEPT | Arithmetic provides simple ways to manipulate numbers. Even with tallies there were two of the basics of arithmetic: addition increased the items on a tally; subtraction removed them—corresponding to, for example, storing and removing bags of grain at a granary. The third operation, division, was also a natural concept, particularly when food had to be divided up. Although easier to perform, multiplication (the reverse of division) is the most sophisticated of the four operations. It may have begun as repeated addition, though it would also have been useful when calculating the area of a piece of land. Division also provides the concept of a fraction— the result of dividing a whole number by another. However not all numbers between the integers are such ratios—irrational numbers, for example, like the square root of 2, could only be represented numerically once decimal fractions were introduced. Early arithmetic involved positive numbers, but the action of removing objects from a collection produces a negative inverse. If we start with 7 apples and end with 2, then there are 5 apples removed, usefully indicated as −5, to distinguish it from apples added. The result was a continuum of numbers from negative to positive, known as the "number line."

DRILL DOWN | As it has only been relatively recently taught in schools, the concept of the number line can seem modern, but it's not. A kind of number line, for example, was involved in the BCE/CE dating system (dating forward and backward from the birth of Christ) popularized by the Venerable Bede in the eighth century. Addition and subtraction become simply moving up and down an imaginary line showing all the numbers in order. Negative numbers helped complete the number line by providing a mirror image of the positive side. If the number line is thought of as a ruler, with the integers providing the marked "notches," fractions fill in the gaps between them.

FOCUS | *The ancient Greeks struggled with fractions. They conceived of fractions as the parts that made up a whole, and considered all fractions, with the exception of $^2/_3$, as $^1/_n$ (if necessary, multiple copies were added together), where* n *is a whole number. A fraction was represented by a letter with a mark to indicate that it was a fraction, making manipulation of them difficult.*

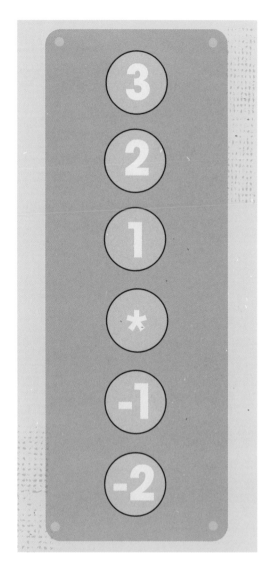

PRIME NUMBERS

THE MAIN CONCEPT | Not all whole numbers behave the same way. There is the distinction between even numbers (divisible by 2) and odd numbers (not divisible by 2). But as numbers get bigger, they can have factors—smaller numbers up to half their value that they can be exactly divided by. (Apart from its own value, a number can't be divisible by a number bigger than half its value.) But some numbers can't be exactly divided by anything other than 1 or themselves—these are prime numbers. So, for example, 2, 3, 5, 7, 11, 13, 17, and 19 are prime, whereas 4, 6, 8, 9, 10, 12, 14, 15, 16, 18, and 20 are not— the second list can all be divided by another whole number. The number 1 seems prime, but we don't count it as one because it breaks the "fundamental theorem of arithmetic," stating that any number can be written as the product (multiplication) of a unique set of other primes. For example, 12 can be written as $2 \times 2 \times 3$. If we allowed 1 in, we could get to 12 using $1 \times 2 \times 2 \times 3$ or $1 \times 1 \times 2 \times 2 \times 3$, or $1 \times 1 \times 1 \times 2 \times 2 \times 3$. As we need a unique set of primes, 1 was ejected. Prime numbers are interesting because of the way that they jump around— there is nothing neatly regular about the sequence as there is in, say, the odd and even numbers. Of the primes, 2 is the only even value, so is also something of a special case. If not a prime itself, a number is called a "composite," for example, $20 = 2 \times 2 \times 5$.

DRILL DOWN | Prime numbers appear less and less frequently as values get bigger. However, more than 2,300 years ago the ancient Greek mathematician Euclid proved with some simple logic that there is an infinite set of prime numbers. He first imagined multiplying together every prime you can think of to produce a value N, then added 1 to that value. Either the new number created ($N+1$) is a prime, or it is made up by multiplying together primes that weren't in the original list, as dividing $N + 1$ by any of the original primes would leave a remainder of 1. So however long your list of primes, there will always be at least one extra.

NUMBERS
Page 22

BASIC ARITHMETIC;
NEGATIVE NUMBERS
Page 26

INFINITY
Page 94

FOCUS | *Prime numbers may seem only of interest to mathematical geeks but are essential to keep us safe on the Internet. When we use a secure connection, generating the encryption key typically involves multiplying together two huge prime numbers. It's easy to multiply them, but almost impossible to deduce the primes from the answer: this hard-to-reverse process is at the heart of one of the most commonly used encryption methods.*

AXIOMS & PROOFS

THE MAIN CONCEPT | We are used to mathematics being very formal and structured, however, very early math was far more pragmatic—more like a cook's recipe. From practical experience, a certain combination of things worked and if they did, they would be repeated and used again. Long before the formal proof of what we now know as the Pythagorean theorem, the Babylonians and other early cultures knew from experience that triangles with a right angle had sides that were in the proportions of 3:4:5. These triangles were very useful in measuring plots of land and designing buildings. It was only after more than 1,000 years of using such "recipe-based" math that the idea of a formal proof began to be used by the ancient Greeks. The idea was to start with a set of axioms (a word that means "that which is self-evident"). These were assumptions that were so obvious that no proof was needed, such as any two points can be joined by a straight line. Proofs were then built up, using logical steps, from the axioms. In the case of the Pythagorean theorem, the Greeks went from simply observing the 3:4:5 ratio to proving that the square of the length of the longest side (the hypotenuse) is equal to the sum of the squares of the other two sides.

DRILL DOWN | So far we've talked about one kind of proof—the kind where, starting with a handful of axioms, mathematicians produced more and more sophisticated theorems, each a collection of arguments that proves a relationship based on axioms and simpler theorems. However, even in ancient Greek times other mechanisms were used for proofs, such as logical arguments or "proof by contradiction." The latter was employed in the proof that the square root of 2 is not a rational fraction. The method involves assuming the result required *is* true—in this case, that there is a rational fraction that equals the square root of 2— and showing that this produces an impossible outcome.

FOCUS | *Many mathematicians dislike proof by exhaustion: trying out every possibility and checking that they behave as expected. However, the four-color theorem, which says that it only takes four colors to make a map where no country has the same color as an adjacent one, was proved this way, working through 1,936 options (later reduced to 1,476) using a computer—the first major theorem of its kind to be proved with this method.*

SERIES

THE MAIN CONCEPT | A series gives the sum of an infinite set of values, where each value is generated from the previous one using a rule. The infinite nature is indicated by stating the first few entries, followed by an ellipsis (. . .). Many series have an infinite sum, for example 1, 1, 1, 1 . . . or even 1, $^1/_2$, $^1/_3$, $^1/_4$ . . . However, some series tend to a finite value. So, for example, the sum of 1, $^1/_2$, $^1/_4$, $^1/_8$ . . . would be 2 for an infinite set. The idea that an infinite series can have a finite sum is challenging: ancient Greek philosopher Zeno illustrated this with the story of Achilles and the tortoise. Achilles runs faster than the tortoise, so gives it a headstart. Achilles soon reaches where the tortoise was when he started, but by then, the tortoise has moved on. This repeats over and over: according to Zeno, Achilles will never catch the tortoise. In practice, though, the distance between them is like the series 1, $^1/_2$, $^1/_4$, $^1/_8$ . . .; there's a finite sum of distances, which Achilles soon passes. Not all series are so simple. The rule can vary, for example, by oscillating—for example, 1, 2, 0, 2, −2, 1, −5 . . . Series proved valuable in the development of calculus (see Chapter 3) and in the calculation of constants.

DRILL DOWN | The familiar constant pi (π), the ratio of a circle's circumference to its diameter, is an example of a constant that can be calculated using a range of series. The more entries in the series that are added, the closer it gets to the exact value of pi, though some series converge on the value faster than others. For example, pi can be calculated using the sum of the series $4/1, -4/3, 4/5, -4/7\ldots$ or π-3 results from the sum of the series $4/(2\times3\times4), -4/(4\times5\times6), 4/(6\times7\times8), -4/(8\times9\times10)\ldots$ Pi can also be calculated by a different type of infinite sequence where the operation is not addition, such as $\pi/2 = (2/1) \times (2/3) \times (4/3) \times (4/5)\ldots$

NUMBERS
Page 22

PI
Page 58

INTEGRAL CALCULUS
Page 100

FOCUS | *Because a series is made up of an infinite set of entries it is susceptible to the paradoxes that arise from infinite calculations. Take the sum of the series 1, −1, 1, −1, 1 . . . By bracketing pairs, the sum is 0: (1 − 1) + (1 − 1) . . . but it also gives a total of 1 if the brackets are shifted: 1 + (−1 + 1) + (−1 + 1) . . .*

POWERS, ROOTS & LOGARITHMS

THE MAIN CONCEPT | Raising a number to a power (called "exponentiation," as the size of the power is the "exponent") is a staple of mathematics. A positive whole number power represents multiplying a number by itself that many times. So, for example, 3^2 is 3×3, while 3^3 is $3 \times 3 \times 3$. Powers turn multiplication into addition. If we multiply 3^2 by 3^4, we add the exponents to get 3^6. This shows that 3^1 is just 3—when we multiply 3^4 by 3 we get 3^5, adding 1 to the exponent. Similarly, dividing involves taking exponents away. So, 3^5 divided by 3^2 is 3^3. But powers are not limited to positive integers. Dividing 3^1 by 3^1, which must give 1, produces 3^0 by subtracting exponents—any number raised to the power 0 gives 1. If we divide 3^0 by 3^1 we get 3^{-1}. This is $1/3$, so 3^{-1} is the inverse of 3^1, and so on for other negative powers. Fractional powers produce "roots." As $3^{1/2} \times 3^{1/2}$ is $3^{1/2+1/2}$, we get 3^1—so $3^{1/2}$ is the square root of 3. Roots have the same hierarchy as powers, with cube roots and so forth. Logarithms, devised in the seventeenth century, came from the observation of the behavior of powers—the way that they can be added and subtracted. Tables of powers were provided to make multiplication and division simpler.

DRILL DOWN | In a logarithm, often shortened to "log," a fixed value (the "base" of the logarithm) is raised to a power to represent a number—the power required to represent a number is its logarithm to that base. So, for example, the log to base 10 of 100 is 2 because $100 = 10^2$. Although base 10 is easiest to get our head around, natural logarithms (shortened to *ln*) are widely used in science, where the base is *e*, a constant of nature with a value around 2.718. This is because natural logarithms make it easy to calculate the way that changing values grow with time. Similarly, logs to base 2 are valuable in computing because of its basis of binary arithmetic.

FOCUS | *The most famous bit of math involving powers is Fermat's Last Theorem, which says you can't have three whole numbers—say x, y, and z—where $x^n + y^n = z^n$ for any integer powers other than 1 or 2. French mathematician Pierre de Fermat claimed to have proved this in 1637, but if he actually did, he never wrote it down. It was proved conclusively by the English mathematician Andrew Wiles in 1994 (see page 75).*

ZERO

THE MAIN CONCEPT | Arguably, one of the biggest steps forward in mathematics involved nothing at all—the introduction of zero. Some early number lines (see page 26) skipped directly from −1 to 1 with nothing in between. We can see that still in our year dating system, which goes directly from 1 BCE to 1 CE. However, the idea of a number representing nothing would change all this. Zeros began as placeholders. When the Sumerians and Babylonians first used their positional numbering system, where the column identified whether a number just meant what it said or was to be multiplied by 60 (60^1), 3600 (60^2), and so on (see page 24), a column that was not in use was just left as a space. This meant that the symbols for, say, 63 and 3603, similar to y yyy and y yyy respectively, were only distinguished by a slightly wider gap between numbers. Eventually it was realized that it would be a good idea to mark an empty column. Now 63 would be y yyy, but 3603 was y \\ yyy. Such placeholder use occurred off and on for centuries. But the big breakthrough came when it was realized that zero could be both a placeholder and the answer to the question "What is (say) 3 − 3?"—the center point on the number line.

DRILL DOWN | Pinning down the arithmetic behavior of zero took some time. It behaves consistently, if oddly, with addition, subtraction, and with multiplication. Adding and subtracting a zero does nothing at all, while multiplying by zero reduces anything else to nothing. But there was confusion over division. The smaller the thing you divide by, the bigger the result. At the extreme of zero, it would seem the result is infinite. And things get even stranger when dividing zero by zero. Some early mathematicians thought the result was zero, others infinite. It is now considered indeterminate—it doesn't have a value, so don't try it (if you see a result of "NaN" on a calculator, it means "not a number").

FOCUS | *Along with Indian/Arabic numerals, the fully functional zero was brought to the West by Fibonacci in his Liber Abaci. He referred to it as "zephirum," which seems to have been an attempt to render the Arabic word* sifr *(which also became "cipher"), a special, egg-shaped empty number. The assumption is that this reached the Arab world from India. Zephirum became zero.*

FIBONACCI NUMBERS

THE MAIN CONCEPT | Just as we separate some whole numbers off as odd or even or prime, there is another subset of the whole positive numbers known as the Fibonacci numbers. These are the numbers in the sequence 1, 1, 2, 3, 5, 8, 13 ... where each number is the sum of the previous two numbers in the sequence (the sequence can also be started with a 0). The name comes from Leonardo of Pisa, the thirteenth-century mathematician better known as Fibonacci. He described the sequence in his book *Liber Abaci*, which also introduced Arabic numerals to Europe. However, these numbers were known up to 1,400 years before in Indian mathematics. Fibonacci showed how the numbers described the way a population of rabbits might increase, but they came to have a wide number of applications. Though the rabbits provided a very artificial example, flowers often average a Fibonacci number of petals, and the numbers are also sometimes reflected in layouts of seed heads and tree branches, simply as a result of the way the structure of a plant builds up. If the numbers are used to create a set of squares with these widths, the result is the spiral shown opposite. And if you divide each number in the sequence by the previous one, it homes in on the "golden ratio" of around 1.618, which many think is artistically and architecturally pleasing (though nobody knows why).

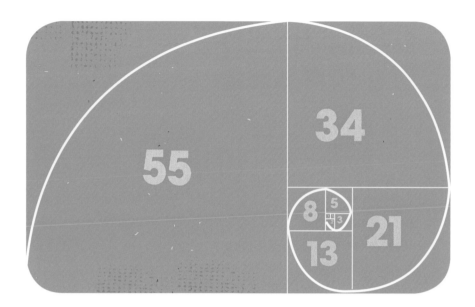

DRILL DOWN | The *Liber Abaci* used Fibonacci numbers to show the number of rabbits in a highly simplified population model. The picture starts with a breeding pair. It takes a month to produce baby rabbits, who are always born in pairs, one male, one female. It also takes a baby rabbit a month to be ready to breed. So, after one month there's still one pair. After two months there's a second pair. After three months the original pair produce a new litter. After four months the first and second pairs produce a litter each. And so on. The result is 1, 1, 2, 3, 5, 8, 13 . . . pairs of rabbits.

FOCUS | *It might appear that the Fibonacci numbers are a very well-explored and simple aspect of number theory, but mathematicians would beg to differ. Ever since 1963, a group known as the Fibonacci Association has published a journal,* Fibonacci Quarterly, *dedicated to the Fibonacci series and associated aspects of mathematics. So far, they haven't run out of things to say.*

NUMBERS
Page 22

**BASIC ARITHMETIC;
NEGATIVE NUMBERS**
Page 26

SERIES
Page 32

IMAGINARY NUMBERS

THE MAIN CONCEPT | One of the most interesting aspects of negative numbers is that when you multiply a negative number by another negative number you get a positive one. So, for example, −2 × −3 = 6. But what, multiplied by itself, makes a negative number? What is a negative number's square root? There is no natural answer, so a purely hypothetical concept, which the philosopher Descartes sarcastically labeled an "imaginary number," was dreamed up. The value i is assigned to the square root of −1. As a result, the square root of any negative number is a multiple of i. For example, the square root of −4 is $2i$ (−2 is approximately $1.41i$, and −3 is $1.73i$). Of itself, an imaginary number might have been an amusing mathematical oddity without any real significance. But such numbers proved extremely useful. When combined with a real number, an imaginary number provides a valuable mechanism for representing points on a two-dimensional plane—and that makes such a "complex number" perfect for dealing with the mathematics of anything with a wavelike behavior, which crops up repeatedly in both physics and engineering. A complex number is represented as, for example, $6 + 2i$, combining the real number 6 and the imaginary number $2i$, and can undergo all the usual operations applied to a number.

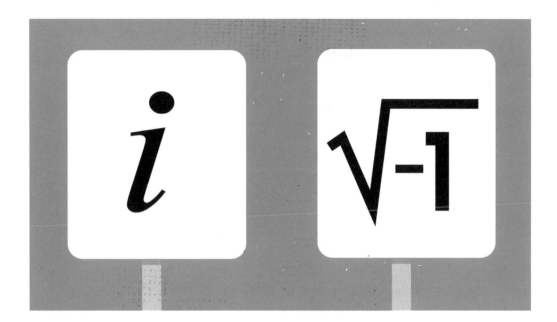

DRILL DOWN | The concept of the number line (see page 26) is central to applying complex numbers to coordinates on a plane. If we imagine a pair of axes, the horizontal axis is the number line of real numbers, centered on zero, with negative numbers to the left of this and positive numbers to the right. The vertical axis is the number line of imaginary numbers, again with zero at the center, positive imaginary numbers above zero, and negative imaginary numbers below zero. Then, for example, the point $3 + 5i$ sits in the top right quadrant of the plane, while $-4 - 6i$ is in the bottom left quadrant. Equations describing the progress of a wave are represented using changing real and imaginary components.

FOCUS | *In the nineteenth century, Irish mathematician William Hamilton introduced a supercharged complex number called a "quaternion." Instead of having a single imaginary element, it had three different imaginary elements. This was designed to handle something that was changing in three dimensions at once, but was largely superseded by a branch of maths known as "vector analysis."*

NUMBERS
Page 22

BASIC ARITHMETIC;
NEGATIVE NUMBERS
Page 26

VECTOR ALGEBRA & CALCULUS
Page 104

MODULAR ARITHMETIC

THE MAIN CONCEPT | When small children are asked to divide 13 by 5, they might say "2, remainder 3," because 5 divides into 13 twice, with 3 left over. As adults we tend to think of division in terms of fractions or decimals—so 13 divided by 5 becomes $^{13}/_5$ or 2.6. However, the child's approach makes sense when dealing with physical objects that can't be split. If we divide 13 pens between 5 people, for example, we can give them 2 each, with 3 left over. Underlying this is modular arithmetic. In modular arithmetic, which only deals with integers, the value cycles back to the start after a maximum, known as the "modulus." So, for example, if the modulus is 5, we count 1, 2, 3, 4, 5, 1, 2, 3, 4, 5, 1, 2, 3 . . . Working with a modulus of 12, for instance, we identify a position on the cycle as, say, 3 modulo 12, or 9 modulo 12 (often shorted to 9 mod 12). Modular arithmetic crops up in everything from clocks to music to ciphers, or codes, all representing cyclical processes. In computing, it's a handy way to test if an integer value is odd or even. If n mod 2 yields a remainder (it will be 1 if it does), then n was odd.

DRILL DOWN | If it's 9 o'clock, 5 hours later it will be 2 o'clock. Normal clock times are modulo 12. Modular arithmetic usually starts at 1 or 0. On the 24-hour clock, 0 and 24 are congruent—they mean the same thing. Similarly, ciphers are often modular. They often use a key, a secret word creating the enciphered text when the number values of each key letter are added to the original text using modular arithmetic. Enciphering DOG with the key CAT we add 3, 1, and 20 to the number values in DOG. This gives GPA. The final letter is G + T or 7 + 20. But the answer is to modulo 26 with 26 letters, so we get 1 or A.

FOCUS | *Modular arithmetic crops up in checksums, where a digit is added to strings of numbers to force a modular result. ISBNs, used to identify books, end with a check digit calculated this way. The old 10-digit ISBNs used modulo 11 arithmetic, using X to represent 10; this has been dropped, but the modulo 10 approach in the new ISBN lets some errors through.*

NUMBERS
Page 22

BASIC ARITHMETIC; NEGATIVE NUMBERS
Page 26

SET THEORY
Page 106

MATHEMATICAL LOGIC

THE MAIN CONCEPT | Logic began in ancient Greek times as a mechanism for exploring reasoning, used to deduce new information from existing relationships. So, for example, if it is true that all dogs have four legs and we know that a particular person has two legs, we can deduce that that person is not a dog. Later, mathematicians devised an approach to logic using the tools of mathematics. This would be turned on its head to define the basic workings of mathematics based on logical premises—building from simple starting points such as set theory (see page 106), and using step-by-step logical processes. The best-known aspect of mathematical logic is Boolean algebra, named after English mathematician George Boole. This uses operators, mathematical tools, to work on the logical values "true" and "false." The simplest operator NOT turns "true" into "false" and "false" into "true." Other operators combine pairs of values. So, for example, AND returns "true" if both inputs are true, and "false" otherwise. Similarly, OR returns "true" if either input is true, and "false" otherwise. These relationships are often portrayed in Venn diagrams. For a long time regarded as interesting but without practical application, Boolean algebra came into its own in the computer, where it is used to manipulate bits—binary digits—substituting 1 for "true" and 0 for "false."

DRILL DOWN | It is easy to spot mathematical logic at play in computers when we try to reduce a large number of items to the relevant ones, particularly in a search. If, for example, I wanted to find a red car, but didn't want a diesel engine, I would be looking for "car AND red AND NOT diesel." However, this logic is also at play at a more fundamental level in all the circuitry that handles data within the computer. This is made up of small parts—originally separate physical components—that are called "logic gates," which carry out Boolean operations on data bits. These gates are now combined by the billion on processor chips.

NUMBER BASES
Page 24
AXIOMS & PROOFS
Page 30
SET THEORY
Page 106

FOCUS | *The most complete attempt to derive the core of mathematics from logic came in the three-volume* Principia Mathematica *by Alfred North Whitehead and Bertrand Russell. Published between 1910 and 1913, it famously took several hundred pages to get to 1 + 1 = 2. The approach took a battering when it was proved by the 25-year-old Austrian mathematician Kurt Gödel that a complete system of mathematics can never be built up this way.*

"[The book of the universe] is written in the language of mathematics, and its characters are triangles, circles, and other geometric figures, without which it is humanly impossible to understand a single word of it."

GALILEO GALILEI, *THE ASSAYER* (1623), TRANSLATED BY STILLMAN DRAKE, *DISCOVERIES AND OPINIONS OF GALILEO* (1957)

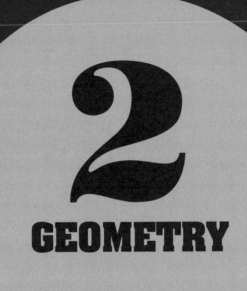

2

GEOMETRY

INTRODUCTION

The word *geometry* means "earth measurement." It is the mathematics of shapes—originally those that could be drawn on a flat surface, but then extended to take in three-dimensional forms and more. Those of us who were taught traditional geometry at school will be familiar with a tedious sequence of theorems, ending in the obligatory QED. (The original Greeks wrote *OEΔ*, the initials for an equivalent term: QED comes from the Latin *quod erat demonstrandum*, meaning "what was to be shown.") But it's a shame that geometry has a boring image, as it is so much more than rote learning. Originally concerned with land measurements, geometry expanded from two-dimensional surfaces to as many dimensions as you would care to consider, and also to related fields such as topology and knot theory.

Euclidean geometry

Going back to basics, the ancient Greek enthusiasm for geometry reflected their visual approach to mathematics—very different from our modern fondness for equations. Instead of manipulating symbols, the Greeks thought in shapes and areas. Even fractions were visual. Rather than say "a quarter," they would use "the fourth part," visualizing four separate parts of the size needed to make up a whole.

Traditional geometry—or "Euclidean geometry," after the fabled ancient Greek mathematician Euclid—begins with a set of axioms (assumptions considered so obvious there is no need to prove them), in turn leading to a set of theorems, which are proved step by step. The best known of all the theorems, the Pythagorean theorem, describes the relationship of side lengths that must exist in a right triangle. The technology might have changed, but even now surveyors make use of Euclidean geometry to deduce distances and angles. But geometry and related subjects, as we've seen, have gone much further. It's strange that the Greeks limited most of their work to the geometry of flat surfaces as so few natural surfaces are flat—the Earth itself, of course, is spherical. It was only in the nineteenth century that the geometry of shapes on curved three-dimensional surfaces became widely studied.

Beyond Euclid

Once the surface a shape is inscribed upon is no longer flat, things get interesting. For example, parallel lines can meet (contradicting Euclid's claim that "two lines are parallel if they only meet at infinity"). If you start from two points on the Equator and head for the North Pole, those lines are parallel at the Equator. But by the time you reach the North Pole, they meet each other. Draw a triangle on the surface of the Earth and Euclid's theorem that the angles of a triangle add up to 180 degrees goes out of the window—the angles add up to more than 180 degrees, while on an indented curved surface, they sum to less than 180 degrees (consider drawing on the inside of a sphere).

Working in three dimensions (four, if you allow for time) is as far as we can sensibly get in the physical world, but despite the difficulty of imagining what is meant by, say, "ten-dimensional space," mathematicians were able to take geometry into the behavior of curves and manifolds. Manifolds belong to a linked branch of geometry known as topology, which concerns the behavior of shapes in multiple dimensions that can be stretched but aren't allowed to be cut or glued. If these seem arbitrary distinctions, all mathematical systems are based on sets of arbitrary rules. (Even Euclidean geometry has its own restrictions separating it from reality.) As far as a topologist is concerned, a ring donut and a teacup with a handle are the same shape—each has a single hole totally surrounded by matter, and the rest is just down to stretching and deforming the one into the other.

Although in physics Einstein's use of non-Euclidean geometry in general relativity and all sorts of manifolds occurring in later physics take the geometric limelight, the biggest impact has been with the concept of fields—aspects of nature that have a value that can change from point to point in space and time—and which have come to dominate physics. And geometry has also found intriguing spinoffs in the nature-reflecting crinkliness and self-similarity of fractals and the oddities of knot theory. This is far more than just triangles.

TIMELINE

EUCLID'S LEGACY
Euclid writes his *Elements of Geometry*, which provides the foundations of both modern geometry and the wider structure of robust mathematical proofs based on a small number of pre-identified axioms. It is used as a textbook for more than 2,000 years.

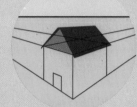

DESCARTES
René Descartes, French philosopher and scientist, publishes *La Géométrie*, as an appendix to his book on the scientific method containing the phrase "I think, therefore I am." In this, far more valuable, appendix, Descartes links geometry and algebra to show the relationship between equations and curves.

c. 2000–1500 BCE

c. 300 BCE

c. 1420 CE

1637

EARLY GEOMETRY
The earliest surviving examples of geometrical working, such as Babylonian clay tablets and the Egyptian Rhind Papyrus, are produced. Evidence of geometric considerations—basic information on triangles and relationships between shapes and areas—goes back around 1,000 years further.

PERSPECTIVE
Italian architect Filippo Brunelleschi makes the first geometric analysis of perspective (though he did not publish details, which would come 20 years later in a book by Leon Battista Alberti). Brunelleschi constructs a reflecting device to enable viewers to compare a view with a perspective painting.

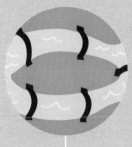

FIELDS

English scientist Michael Faraday coins the term "field" to describe what he has previously referred to as "lines of force"—a mathematical entity with a value at every point in time and space that can be used to describe the influence of electromagnetism: much of modern physics is based on fields.

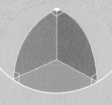

FRACTALS

French mathematician Benoit Mandelbrot publishes "How Long Is the Coast of Britain?," which pulls together thinking on what he will later describe as "fractals" and will lead in 1979 to the introduction of the Mandelbrot set.

1736   **1845**   **1854**   **1967**

GRAPH THEORY

The Swiss mathematician Leonhard Euler publishes what is arguably the first application of "graph theory," applied to the "Seven Bridges of Königsberg" problem—devising a walk across each of the city's bridges without ever crossing any given bridge twice.

NON-EUCLIDEAN GEOMETRY

In the early nineteenth century, Carl Friedrich Gauss, Johann Bolyai, and Nikolai Lobachevsky experiment with versions of geometry where parallel lines meet, moving away from flat surfaces (non-Euclidean geometry). By 1854, Bernhard Riemann has taken an alternative approach to the subject in relation to smooth surfaces.

BIOGRAPHIES

EUCLID (c. 325 BCE–c. 265 BCE)

Euclid was a Greek mathematician who pulled together the scattered work on geometry undertaken in the ancient world and turned it into a refined, structured whole, starting with simple assumptions and building up increasingly complex proofs. Very little is known about him biographically; details about his life, such as his being born in Tyre, seem to be fictional additions by later authors. It has even been suggested that Euclid did not exist as an individual, but rather that his output was that of a group of philosophers making use of a fake name. (This certainly happened in the twentieth century, when a group of mathematicians wrote under the pseudonym Nicolas Bourbaki.) If this was the case, it's likely that Euclid was named for the earlier philosopher Euclid of Megara. Many ancient Greek authors simply referred to "the author of *The Elements*," the 13-volume book on geometry (with a bit of other mathematics) that was Euclid's famous work. It appears to have been written in Alexandria, and he is sometimes referred to as Euclid of Alexandria accordingly.

GASPARD MONGE (1746–1818)

Born in Beaune in Eastern France, Gaspard Monge would make significant contributions to two areas of geometry. After mathematical and physics training at the Collège de la Trinité in Lyon, he became a draftsman at a military engineering school in Charleville-Mézières. He went on to become a professor of mathematics there. Monge became a significant academic figure in the French Revolution. Before he became more of an administrator, his experience as a draftsman stood him in good stead in developing what is now known as "descriptive geometry"—the mathematical procedures needed to project three-dimensional objects onto two dimensions to enable engineering drawings and other plans to be drawn up. He started to develop this while still working as a draftsman, when asked to produce the plan for a fortification that made it impossible for an enemy to see or fire at a military position. Monge's more sophisticated contribution to geometry was the branch known as "differential geometry." Again, he was dealing with three-dimensional shapes, but here it was the use of calculus and algebra to describe the behavior of curves and shapes in a three-dimensional environment. Monge died in Paris, aged 72.

HERMANN MINKOWSKI (1864–1909)

Born in Aleksotas (then in Russia, now in Lithuania), Hermann Minkowski moved with his family to Königsberg (then in Germany, now Kaliningrad in Russia) when he was a child, so he is usually regarded as a German mathematician. Specifically, he was the mathematician who persuaded Albert Einstein to accept a more geometrical view of the world in thinking of space and time as a unified entity: space-time. Minkowski lectured at a number of universities, including the ETH in Zurich, where Einstein was one of his students. However, it was after Minkowski moved on to Göttingen that he worked on the mathematics of space-time and devised the Minkowski diagram, still the main method of representing space-time events. Geometrically his most significant work was on the geometric properties of multidimensional space, including the four-dimensional space-time, and the "geometry of numbers," which links number theory and multidimensional spaces. Göttingen was Minkowski's final teaching post, where he worked with the greatest German mathematician of that period, David Hilbert. Sadly, Minkowski was only 44 when he died of a ruptured appendix.

CHRISTIAN FELIX KLEIN (1849–1925)

Christian Felix Klein was born in Düsseldorf in Germany. He studied at Bonn, originally intending to work in physics, but became fascinated by mathematics, which would take him as lecturer and then professor to Erlangen, Munich, Leipzig (where his chair was explicitly in geometry), and Göttingen. One of his most significant achievements in the field was not directly in mathematics, but as editor of the journal *Mathematische Annalen*, which had been very much in the shadow of *Crelle's Journal*, but became a leading mathematical publication. Klein's dissertation was on the geometry of lines in a plane, and geometry would continue to be the central focus of his output. He carried forward work on non-Euclidean geometry, and, in his so-called Erlangen Program, linked geometry to "group theory" and aspects of symmetry that would become important in twentieth-century physics. He also worked in complex analysis— calculus for complex numbers. To lovers of recreational math, he is best known as the creator of the Klein bottle, a bottle that, like a Möbius strip, has only one side—though it can only be made in three dimensions as a projection of a four-dimensional object.

EUCLIDEAN GEOMETRY

THE MAIN CONCEPT | Euclidean geometry primarily deals with shapes on a flat plane. The mechanism Euclid used in his *Elements* was to start with five simple axioms ("postulates")—statements assumed to be self-evident about lines, angles, and circles—and build a series of proofs on these. These are: a straight line can be drawn between any two points; any straight line can be extended indefinitely; a circle can be defined by its center point and its radius; all right angles are equal; and a fifth, the "parallel postulate," which states that if the angle between two lines is less than 180 degrees they will meet (often phrased as "two lines are parallel if they only meet at infinity"). With the axioms established, Euclid proved a series of theorems—each a statement about the behavior of geometric shapes. The theorems of Euclidean geometry build on one another to prove, for example, that the angles of a triangle add up to 180 degrees, or that if each of two sides on a pair of triangles are of equal length and the angle between them is the same, the triangles are identical, known to geometers as "congruent." Other triangles can be proved to be "similar" to one another, where the same triangle is scaled up or down. The best known of all the theorems, the "Pythagorean theorem," describes the relationship of side lengths that must exist in a right triangle.

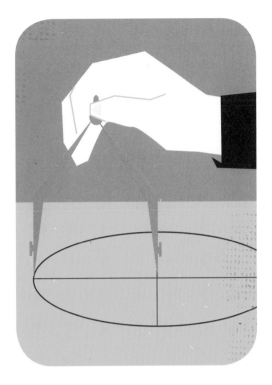

DRILL DOWN | Although classical geometry is sometimes thought of as being all about shapes on a flat, two-dimensional surface, Euclid did give some consideration to three dimensions. One topic of considerable interest in early times was the Platonic solids—three-dimensional shapes that could be made from sides of identical regular flat shapes. There were only five of these: the tetrahedron, octahedron, and icosahedron with 4, 8, and 20 triangular faces; the cube with 6 square faces; and the dodecahedron with 12 pentagonal faces. There is something appealing about the regularity of these solids and some ancient Greeks considered them linked to the five elements of earth, air, fire, water, and the quintessence, or ether.

FOCUS | *Seventeenth-century German astronomer Johannes Kepler believed that there was a relationship between the five Platonic solids and the six known planets: Mercury, Venus, Earth, Mars, Jupiter, and Saturn. He envisaged a structure of spheres, separated by the Platonic solids, producing the scale of the solar system. The concept had no scientific basis: it was an early theory based on "beauty," a concept physicists still pursue.*

TRIGONOMETRY

THE MAIN CONCEPT | The relationship between angles and the length of sides of triangles is a natural extension of geometry, called "trigonometry," meaning "triangle measurement." Thinking of one of the angles in a right triangle, the "sine" of that angle is the ratio of the length of the opposite side to that of the hypotenuse (longest side). The "cosine" is the same measure but using the ratio of the third side to the hypotenuse. And the "tangent" is the ratio of the opposite side to the third side (you may be familiar with the "SOH CAH TOA" method of remembering this). These simple "functions" have proved useful in surveying, navigation (for example, with sextants), and astronomy. Trigonometry also enables triangulation—deducing distances by measuring the angles of triangles between known positions and a new point, which proved transformational for accurate map making. However, the triangular definition of the trigonometric functions was limiting as it could only deal with angles up to 90 degrees. An extended definition imagines the triangle drawn in a circle with the right angle at the center. The hypotenuse is the radius and can be swept around the circle, giving values that oscillate periodically. This approach produces an alternative unit of angle, the "radian." Trigonometry was also extended out of the plane, using a sphere rather than a circle as the base.

DRILL DOWN | Once trigonometric functions were measured using a circle, it became possible to think of the size of an angle as the amount by which the hypotenuse of the triangle has swept around the circumference of a circle. Based on a circle with a radius of 1, the circumference is 2π—so a rotation of 360 degrees is considered 2π radians, the name given to this new unit. This approach proved invaluable when trigonometry was combined with calculus, as using the radian measure produced simpler results. It is also a more mathematically rigorous unit than the culturally influenced degrees. Radians are now the standard scientific unit for the measurement of angles.

FOCUS | *In many countries, high points in the landscape are marked with different styles of concrete pillars, usually topped with a metal plate. Known as "triangulation stations" in the US and "trig points" in the UK, these are mounts for theodolites, optical instruments used to enable trigonometric measurements. They have largely been made redundant by GPS satellite navigation systems.*

PI

THE MAIN CONCEPT | A handful of universal constants are familiar to most of us—and leading the pack is pi (π), the ratio of a circle's circumference to its diameter, named using the first letter of the Greek word for circumference. Approximate values of pi were known at least 4,000 years ago. Later, the ancient Greek mathematician Archimedes came up with a clever mechanism for setting limits on pi—he drew a regular polygon that just surrounded a circle and a smaller similar polygon that just fitted inside the circle. The length of the circle's circumference had to be in between the perimeters of the two polygons. The more sides the polygons had, the more accurate the calculation (as in the illustration opposite). Archimedes used a pair of 96-sided polygons to show that pi lay between $223/71$ and $22/7$ (3.1408 and 3.1429). The number is both "irrational" and "transcendental," meaning effectively that it can't be calculated exactly by a finite mechanism. Far more decimal places were pinned down later by calculating parts of infinite series that sum to pi. It has now been calculated to quadrillions of decimal places. Pi obviously has a use in geometry, but it also turns up in many aspects of physics, from vibrating strings to the uncertainty principle, in the normal distribution of statistics, and in many more places.

DRILL DOWN | Pi appears where it's least expected. Because of its occurrence in probability distributions, a cocktail stick and a set of parallel lines (floorboards will often do) can be used to calculate an approximate value of pi. Repeatedly drop the stick onto an area divided up with lines that are farther apart than the length of the stick. An approximate value for pi is $2ln/wx$, where l is the length of the stick, n the number of drops, w the width of the line gaps, and x the number of sticks that cross a line. The greater the number of drops, the closer the value will get to pi.

FOCUS | *Ever since ancient times, geometers have tried to "square the circle"— devise a way to produce a square with the same area as a circle using only the geometry tools of a straightedge and compasses. The ancient Greeks even had a name for people who tried to do this:* tetragonidzeins. *Sadly, because pi is transcendental, squaring the circle is impossible.*

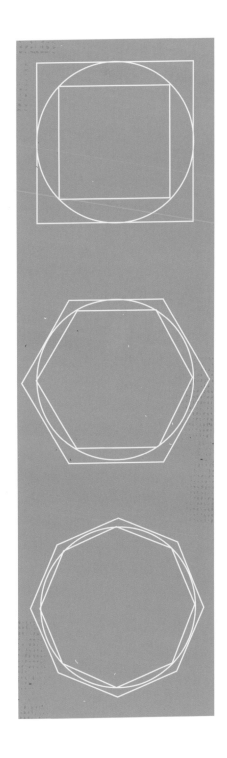

CONIC SECTIONS

THE MAIN CONCEPT | Conic sections, which have been studied *ad nauseam* through mathematical history, are simply the shapes obtained by cutting a slice from a cone. Although an arbitrary-sounding concept (why not sausage sections or Gummy bear–shaped sections, say?), it proves valuable because the cone is unique in the range of useful shapes produced this way. Slice it parallel to its base and you get a circle. Cut it across the cone at an angle to the base but staying within the cone and you get an ellipse. Make the same cut but slicing through the base gives you a parabola. And making a similar cut but with all of the slice on one side of the vertex gives a hyperbola. (Strictly you need two cones, point to point, to get both pieces of this "bipartite" curve.) When it comes to applications, all the sections have metaphorical uses—we speak of a circular argument and an elliptical expression, parables are parabolic, and when we come up with hyperbole it is excessive, just as the hyperbola exceeds the bounds of the cone. More practically, all the conic sections turn up in astronomy, and most are relevant in areas of physics and engineering, from mirror shapes to the design of gears.

DRILL DOWN | The topic where conic sections probably have the broadest application is in orbits. For a long time, planetary orbits were assumed to be circles—the "perfect" shape—and many orbits are in fact nearly circular. But German astronomer Johannes Kepler showed observationally that planets follow elliptical orbits, while Isaac Newton proved mathematically that a gravitational inverse square law—where the attraction falls off with the square of the distance between the bodies— produces an ellipse. Not every object stays in continuous orbit, though—some comets, for example, dip into the solar system and fly out again. Depending on whether or not they go around the Sun, their orbits follow either a parabola or a hyperbola.

FOCUS | *The practical applications of conic sections tend to be of particular value to specialists, but a treatise on them gave us an iconic symbol. In his 1655* Tract on Conic Sections, *Oxford mathematician John Wallis wrote, "let the symbol ∞ represent Infinity." He never said why he chose this distinctive shape, though some have suggested it's the next simplest continuous loop after 0.*

GEOMETRY & ASTRONOMY

THE MAIN CONCEPT | Astronomy and mathematics have gone hand in hand since humans began studying the heavens—so much so that astronomy was counted part of mathematics rather than the natural sciences until well into the nineteenth century. Keeping track of the motion of the planets was difficult, and sometimes even weird, particularly in pre-Copernican days—before Nicolaus Copernicus proposed that the Sun, not the Earth, sat motionless at the center of the universe, with the planets orbiting around it. Sometimes outer planets appeared to reverse their direction of travel. We now know this is because the Earth's orbit means we see, for example, Mars's orbit from a changing viewpoint. But when it was thought that everything traveled round the Earth, complex geometric patterns known as "epicycles," where smaller circular motions traveled around larger orbits, were required. Although Newton used calculus to work on his gravitational theory, his masterpiece on gravity and forces, the *Principia*, almost entirely employs geometry and the properties of conic sections in its arguments. Geometry would also be used to calculate astronomical distances, from Archimedes's early attempt to calculate the size of the universe to modern-day parallax methods. It is also required in telescope design.

DRILL DOWN | Hold a finger in front of your face, then open and close each eye in turn. The finger appears to move from side to side. Do so with an object farther away, and it seems to move less. This is parallax. By measuring how much the object appears to move, and knowing the distance between our eyes, we can use simple trigonometry to work out the distance to the object. Astronomers often use separate telescopes this way, but a more powerful parallax technique uses the same view six months apart, when the Earth will be on the opposite side of its orbit—the equivalent of using eyes separated by 190,000,000 miles (300,000,000 kilometers).

FOCUS | *The strangest book Archimedes wrote was called* The Sand Reckoner. *In it, he estimated the number of grains of sand it would take to fill the universe. His intention was to show how the limited Greek number system could be indefinitely extended. But to do so he had to employ several geometric techniques to estimate the size of the universe before considering filling it.*

PERSPECTIVE & PROJECTION

THE MAIN CONCEPT | Very old paintings have a strange, flat appearance—there is no depth because the painter has not taken into account the geometric effect of perspective. Put simply, the farther away things are, the smaller they appear to be. To get an image correctly in perspective involves drawing (literally or mentally) a series of perspective lines from the viewer's position into the distance. Parallel lines in the actual world converge at distant points known as "vanishing points." The idea of perspective was so novel when architect and artist Filippo Brunelleschi demonstrated it in the early fifteenth century that he needed a visual aid. This was a mirror image of the Florence Baptistery, painted with perspective on the back of a board, with a hole through the vanishing point; as a nice detail, Brunelleschi made the sky of polished silver, so it reflected the clouds. Viewers looked at the real baptistery through the hole on the reverse side of the board, then a mirror was used to switch their view to the painted version. Perspective views are part of graphical projection, where a three-dimensional object is projected onto a two-dimensional surface. The shape of a cube, for example, alters significantly when drawn from different viewpoints. Projection is essential both for providing maps of the Earth on flat paper and in technical and architectural drawings.

DRILL DOWN | We are so used to projected maps of the Earth that it is easy to forget that what is seen cannot be 100 percent accurate. Something has to be lost in going from a spherical surface to a flat map. The familiar map of the world is a variant of cartographer Gerardus Mercator's 1569 "Mercator projection," which projected the features of the Earth onto a cylinder surrounding the equator and was then unfurled. The result is beneficial for nautical navigation but provides a distorted view of the areas of countries: areas are magnified as you move away from the equator. For example, Greenland looks much bigger than India; in reality, they are very similar in area.

FOCUS | *One effect of perspective is that the farther away an illuminated object is, the dimmer it appears to be. This effect is used by astronomers to estimate distances using "standard candles." These are particular types of astronomical object that appear to have consistent brightness. The first standard candles were stars known as Cepheid variables—identified in 1912 by astronomer Henrietta Swan Leavitt as having a regular cycle of brightness.*

GEOMETRY & ASTRONOMY
Page 62
GRAPH THEORY
Page 68
NON-EUCLIDEAN GEOMETRY
Page 70

CARTESIAN COORDINATES

THE MAIN CONCEPT | In the early days, geometry was considered totally different from arithmetic—as different as art and music are to each other. It wasn't just that there was no obvious way of connecting them, they were simply considered totally different things. By the seventeenth century, though, French philosopher and mathematician René Descartes was the first to widely publish a link between the two, although others—notably French mathematicians Pierre de Fermat and Nicole Oresme—achieved similar results. The essential step was what became known as "Cartesian coordinates," after Descartes. The system used today (incorporating a refinement made to Descartes's original, see opposite) involves using a pair of number lines, one horizontal and one vertical, crossing at zero. By convention, the horizontal line is the "x axis," and the vertical line the "y axis." Then a point on a two-dimensional plane, for example, 1 unit to the right on the x axis, and 2 units up the y axis, can represent variables x and y with the values 1 and 2 respectively. In principle, there can be more axes at right angles to the existing pair—dealing with as many dimensions as a mathematician wants to play with. The real power of the system is that an algebraic equation, such as $y = x^2 + 3x + 2$ now became simply a curve, mapped out with Cartesian coordinates (a parabola in this case).

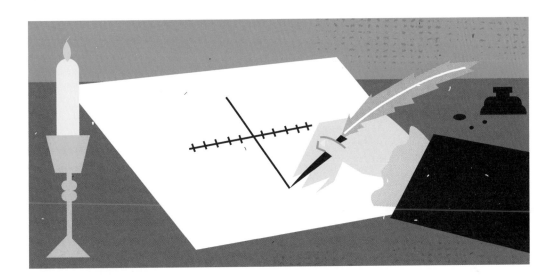

DRILL DOWN | The ability to relate equations and curves meant that Cartesian coordinates were far more than simply a way to give an identifier to a position on the plane. It made possible what would become known as "analytic geometry," where the results of different equations could be derived from curves (or later, multidimensional shapes). When Newton and Leibniz came up with calculus, this relationship between equations and curves or shapes would be fundamental to the development of what proved to be one of the most versatile and powerful tools of mathematics. Arguably, the introduction of Cartesian coordinates was the point at which mathematics became the essential tool for physics.

FOCUS | *Although Descartes is credited with inventing the Cartesian coordinate system, unlike the one we use today, his original version had only a single number line, the x axis. The other value was simply represented by a measurement away from it in the up or down direction, rather than having a second axis. This would be added a decade or so later by other mathematicians, making Descartes's concept complete.*

LINEAR ALGEBRA
Page 92

DIFFERENTIAL CALCULUS
Page 98

INTEGRAL CALCULUS
Page 100

GRAPH THEORY

THE MAIN CONCEPT | Topology (see page 72) involves reducing shapes to flexible equivalents, but a subset of topology takes things even further, leaving the bare skeleton of the original object. Where most of us think of a "graph" as an alternative term for a chart, to mathematicians it is a visual representation of this cut-down form. Graph theory converts more complex objects—both physical or virtual—into dots (known as "nodes" or "vertices") connected with "lines" (or "edges"). Such graphs can refer to shapes, corresponding to the vertices and edges of that shape, but are also used to study networks of relationships. Although not formally identified as such, graph theory is generally said to have originated with German mathematician Leonhard Euler's 1736 analysis of the "Seven Bridges of Königsberg" problem (see opposite), though it was not formally established until the late nineteenth century. Examples of graph theory include the types of family tree that feature each person as a node, with the lines representing familial links. Similarly, the trees and cladistic diagrams of biology are graphs. Sociologists use graphs to describe social networks, while in computing they are widely used both to analyze a communications network—the Internet, for example—and the organization of data. Graph theory has proved useful wherever there is a structure that involves linkages between data points.

DRILL DOWN | The Seven Bridges of Königsberg problem (below) involved finding a route across each of that city's seven bridges without crossing a bridge twice. Leonhard Euler approached the challenge by effectively introducing graph theory—ignoring geography and turning land masses into nodes, and bridges into lines. Leaving aside the first and last node, every other node had to have an even number of attached lines. This was because a line could only be used once, so these nodes needed the same number of entrance and exit lines, otherwise the route would end at that node. But in Königsberg, each actual land mass had an odd number of bridges leading to it; the route didn't exist.

EUCLIDEAN GEOMETRY
Page 54

KNOT THEORY
Page 76

SET THEORY
Page 106

FOCUS | *A popular concept derived from graph theory is six degrees of separation— the idea that everyone in the world is a maximum of six contacts away from everyone else. The number six is not rigorously scientific, emerging from a piece of 1920s fiction and ad hoc experiments involving forwarding letters in the 1960s, but we do live in a strongly connected world.*

NON-EUCLIDEAN GEOMETRY

THE MAIN CONCEPT | Although the ancient Greeks extended geometry beyond a flat plane to take in simple three-dimensional forms such as a sphere, cone, and Platonic solids (see page 55), it does not seem to have occurred to them to extend the theorems of geometry, such as the sum of the angles of a triangle, to deal with the familiar three-dimensional space that we occupy. It's perfectly possible to do geometry on a curved space in three dimensions, for example, such as the surface of the Earth. Here angles of a triangle add up to more than 180 degrees, while the shortest distance between two points, a straight line on a flat surface, becomes a "great circle"—a line connecting the two points that follows a circumference of the sphere. Simple non-Euclidean geometries are either elliptical—where the surface curves as an ellipse (including the special case of a sphere)—or hyperbolic, where the surface is the interior of the curves of a hyperbola and triangles have angles that add up to less than 180 degrees. This is only the start though—there is no reason mathematically to be limited to three dimensions. And more sophisticated geometries exist, such as Riemannian geometry, which allows for non-regular surfaces, and where the main requirement on the curved space is that it does not vary discontinuously.

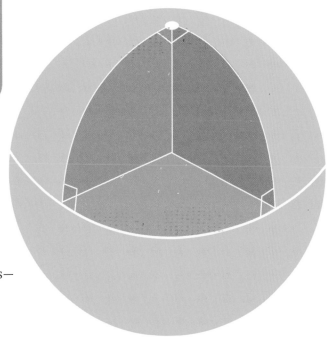

DRILL DOWN | The use of three dimensions in geometry seems perfectly natural because we are familiar with orientation and measurement in three-dimensional space. But what about further dimensions? Even the natural world demands we consider four dimensions—with time as the fourth—while mathematicians deal with imagined spaces containing thousands of dimensions. No one can visualize what a more-than-three-dimensional object looks like, but there are some commonly used four-dimensional shapes, such as the tesseract, or hypercube, which has a total of eight cubic sides. In practice, though the mathematics may be working in many dimensions, illustrations are projected onto two or three dimensions, often by selecting the most relevant to observe.

FOCUS | *The most famous application of non-Euclidean geometry is Einstein's general theory of relativity, which is our best theory of gravitation. It was nearly Hilbert's general relativity. Einstein was struggling with the mathematics of curved space and mathematician David Hilbert set out to produce his own equations. They were ready to be published before Einstein's, but the last-minute discovery of an error held Hilbert up.*

TOPOLOGY

THE MAIN CONCEPT | The word "topology" literally means the "science of place." It's the mathematics of surfaces and objects that has total disregard for size and shape—it is just concerned with how a substance (real or imagined) is either continuous or broken. Two shapes are topologically identical if one can be deformed into the other without tearing it, just stretching it or compressing it as required; it has sometimes been described as "qualitative geometry." So, for example, thinking of pasta shapes, spiral fusilli are topologically identical to spaghetti—neither has any openings—while penne and cannelloni are topologically identical, with a single hole. Topology is also concerned with concepts such as surfaces and edges. So, for example, a ribbon that makes a circular loop has two sides and edges (being a perfect, mathematical object with no thickness), while a looped ribbon with a twist in it—a so-called Möbius strip—has just one side and one edge. There is no way to transform one into the other by simply stretching it. Topology has also been extended to include the equivalent properties of sets. Apart from its mathematical interest, topology is relevant in physics and engineering where surfaces and manifolds are involved, while the linked area of topology known as knot theory (see page 76) is useful both in physics and biology.

DRILL DOWN | Although it's possible to get an overview of topology by thinking of what can be manipulated in a sheet of infinitely stretchable and squashable rubber, there are some definitions required. Most notably there is a divide on whether it is possible to so compress a part that sticks out until it is no longer there. Topology that assumes this is possible is known as "homotopic." This approach regards, for example, the letters O, D, and R to be identical, as the legs of the R could be squashed to nothing. If the assumption is that there will always be some residual bumpiness, the approach is "homeomorphic." Here O and D are identical, but R isn't.

FOCUS | *Topology was a relative latecomer as a mathematical discipline, with the term "topologie" first used in German in 1847 by the mathematician Johann Listing, and not gaining its English form until the 1880s. We perhaps should call the familiar Möbius strip a Möbius-Listing strip as Listing came up with the concept in the same year as August Möbius, and went on to study more complex twisted structures.*

GRAPH THEORY
Page 68

KNOT THEORY
Page 76

SET THEORY
Page 106

ALGEBRAIC GEOMETRY & FIELDS

THE MAIN CONCEPT | "Algebraic geometry" sounds like another term for the links between algebra and geometry formalized by Descartes (see page 66)—and its starting point is a subset of that, focusing on the points at which a polynomial (what we'd think of as one side of an equation, with a mix of constants and variables) hits zeroes. But algebraic geometry leads on to the more general study of the way the solutions of sets of polynomials vary in a geometric fashion. This has become a major area of study in mathematics, whether for abstract mathematical benefit—the proof of Fermat's Last Theorem made use of algebraic geometry techniques (see opposite)—through to modern physics, which since the nineteenth century has been increasingly dependent on such solutions, especially since Einstein's general theory of relativity, and continuing with concepts such as string theory. The fundamental change in physics was the idea of a "field," originated by Michael Faraday and made mathematical by Scottish physicist James Clerk Maxwell. At its simplest, a field is just a number space—it's a mathematical construct that has a value at points in space and time. So, for example, the altitude of every point on the surface of the Earth is a field. Farady used lines to show the strength of a field, such as the Earth's magnetic field, where the field is stronger the closer the lines are together.

DRILL DOWN | Fields were first used on electricity and magnetism, but have been extended to everything from gravity to the Higgs field, which gives fundamental particles their mass. Maxwell took fields from a qualitative idea to a quantitative mechanism, forging a bridge between maths and physics. An essential contribution was the realization that the values at points in the field could represent "vectors"—values with both a size and direction—as well as "scalars" (just numbers) and the development of mathematics to deal with their changes. Some physicists consider the entire universe, sometimes referred to as "the bulk," to be a collection of interacting fields. Working with fields in this way has strong parallels with algebraic geometry.

FOCUS | *Pierre de Fermat's proof of his "last theorem" amounted to a few words scribbled in a margin, claiming there wasn't room to write it there. He was unaware of the mathematics that Andrew Wiles would use between 1993 and 1995 in a pair of papers totaling 129 pages in length, and making use of number theory and the algebraic geometry of elliptic curves to crack the 350-year-old problem.*

KNOT THEORY

THE MAIN CONCEPT | Mathematicians have long been fascinated by knots—though being mathematicians, their version of a knot bears limited resemblance to the real thing: one of the requirements for a mathematical knot is that the "string" the knot is in must have no ends. Also, like the lines of classical geometry, these strings are strictly one-dimensional, having no thickness. Knot theory is an important part of topology (see page 72), and as always in topology, two examples are considered identical if it's possible to transform one into the other without tearing—in this case, without cutting that continuous loop. There is a hierarchy of knots, starting with the open loop (entertainingly known as the "unknot") and adding in more and more loops—so the next form up, for example, is the "trefoil," where the string overlaps at three points to produce three lobes. The mathematical study of knots may seem abstract, but like so many areas of pure math it has proved to have value. Knots, like molecules, can have chirality (handedness) and symmetry, and the parallels between knot theory and, for example, the structure of DNA and the folding of proteins has proved valuable. Similarly, physicists have found the mathematics needed to deal with transforming knot structures useful in both statistical mechanics and quantum computing.

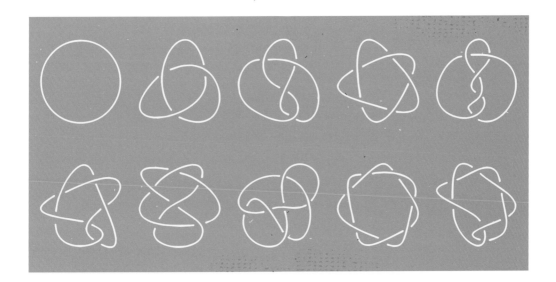

DRILL DOWN | When knots were first studied in the nineteenth century, a leading proponent was Peter Guthrie Tait, a lifetime friend of James Clerk Maxwell (see page 74). Tait was inspired by an (incorrect) theory by William Thomson, later Lord Kelvin, suggesting that atoms were knots in the ether—the rarefied substance assumed to fill all of space, and which light waves passed through. Tait compiled a table of "prime" (non-compound) knots with up to ten crossings, of which there proved to be 165. Tables have now reached at least 16 crossings, producing an impressive 1,388,705 different knots. Though knot theory became part of topology, its origins were therefore, in part, a failed physical theory.

FOCUS | *An unlikely sounding feature of knots is their "writhe," a measure of how coiled up the string is. The writhe is, loosely, the number of positive crossings in the knot minus the number of negative, where a positive crossing is one where the string under the crossing goes from right to left. It's possible to change writhe without changing the topology by twisting the string.*

FRACTALS

THE MAIN CONCEPT | Traditional geometry has largely focused on regular, smoothly shaped objects—but the natural world is crinkly. Fractals are mathematical forms more like real-world shapes. This often involves having structures that naturally scale up in a consistent way. If you zoom into a part of a fractal, it will look similar to the whole—a concept known as being self-similar. Again, this is a common feature of the natural world where, for example, the branch of a tree often resembles a scaled-down version of the whole tree. Fractals often display interesting behavior. Arguably the earliest example, the Koch snowflake, has an infinite perimeter yet can be enclosed in a finite area. It was soon followed by the Sierpinski triangle (or gasket, illustrated opposite), an intricate pattern that manages to have no area at all. The best-known exponent of fractals was French mathematician Benoit Mandelbrot, who gave them their name in 1975 and popularized fractals as graphic art with computer-generated images such as the Mandebrot set, based on iteratively evolving, relatively simple mathematical formulae. As well as being used to generate CGI landscapes, an early hope for fractal technology was highly compressed images. This was extremely effective, but was overtaken by hardware developments. However, fractal methods are used to structure data in a number of analytical methods.

DRILL DOWN | By the late 1980s, computers were handling increasingly large images with expanding color palettes—images were growing faster than the disk technology to store them. The English mathematician Michael Barnsley devised a method using fractal compression that looked for fractal style similarities, so was particularly good on natural images. The compression was effective, but by the early 1990s the simpler JPEG compression standard was available and disk storage grew fast enough to cope with most requirements. Although fractal compression could render some images far smaller than JPEG (an image could be compressed into an "equation"), it was much more processor intensive. There were applications of the technology, from Microsoft's *Encarta* encyclopedia to some video games, but fractal compression has yet to become a mainstream approach.

FOCUS | *Mandelbrot's breakthrough 1967 paper "How Long is the Coast of Britain? Statistical Self-Similarity and Fractional Dimension" analyzed the coastline paradox—an oddity not noted until the twentieth century. It's impossible to say how long a coastline is, as the shorter the units you measure it in, the more you can dip into crinkles, giving a longer result. The value for Britain varies by hundreds of miles, depending on the measure.*

"The study of mathematics, like the Nile, begins in minuteness, but ends in magnificence."

CHARLES COLTON,
*LACON: OR MANY THINGS IN FEW WORDS;
ADDRESSED TO THOSE WHO THINK* (1820)

3

ALGEBRA & CALCULUS

INTRODUCTION

There's an infamous student's complaint, "When will I ever need to use algebra?" Leaving aside the many jobs where it can be directly useful, learning algebra teaches us far more than a way of determining the value of the mysterious x; it's about understanding how to manipulate data in a logical way—something that is beneficial to everyone. Algebra provides a wonderful mechanism for generalization. Where arithmetic deals with explicit numbers, algebra provides us with "variables" such as x and y, which act as containers into which we can plug a whole range of numbers to get a new result. In effect, an "algebraic equation" is a simple kind of algorithm for manipulating data. Yet it didn't start in such a sophisticated form.

The development of algebra

Algebra arguably began with the third-century Greek philosopher Diophantus. He worked with algebraic expressions, though lacking the symbols for operators, he would have rendered, say, $5x^4 + 2x^3 - 4x^2 + 3x - 9$ as something like SS5 C2 x3 M S4 u9. (Of course his version would have been in Greek letters, but this example uses S for square, C for cube, x for unknown, M for minus, and u for unit). However, all of his work was one-off—he didn't generalize the way modern algebra does.

The book that brought algebra to medieval Europe, Al-Khwarizmi's *Hisab al-jabr wa'l muqābalah*, was even less like modern algebra in that it only presented the problems to be dealt with, and the workings, in word form. It wasn't until the sixteenth century that symbols like +, –, and = began to be introduced and algebraic equations could be relatively easily manipulated. But Al-Khwarizmi did show the all-important generalization of solutions. His simple equations were just the start of taking mathematics to the next level, and few contributions to mathematics have proved as powerful as one of the next major developments, calculus.

Calculus

Calculus makes it possible to calculate the outcome from constantly changing values, to work out the areas and volumes of interestingly shaped objects, and far more. Predecessors to calculus date back to the ancient Greeks, who, for example, calculated the area of a circle by dividing it up into more and more triangular-shaped segments. But the full mathematical approach was devised seemingly independently by Isaac Newton and Gottfried Leibniz in the seventeenth century.

Although the seventeenth-century version of calculus made some dubious assumptions that had to be ironed out later to be made mathematically precise, it worked wonderfully in analyzing the mathematics of change and space (it is now more often refered to as "analysis"). Much scientific calculation, particularly in physics, depends to a great extent on the use of calculus.

Sets, groups, and infinity

Equally important to the ability to use mathematics and understand its nature was the introduction of "sets." These crop up as the basis of arithmetic, and took on a whole new level of capability when their definition was expanded to include a special type of set known as a "group," which has a mechanism for generating members of the group from other members. Some groups would prove of particular importance to dealing with symmetry, a mathematical concept central to understanding nature. The use of symmetry in science began with one of the greatest female mathematicians of history, Emmy Noether, who discovered that there was a direct relationship between symmetry and the conservation laws in physics—one was not possible without the other.

Sets also cropped up in the study of the most mind-boggling aspect of mathematics, infinity. Ancient Greek philosophers had contemplated what happens as we count to greater and greater numbers. However, the Greek word for infinity, *apeiron*, has connotations of chaos and disorder. It was not thought of as anything that could have real mathematical significance, but rather something that could be envisaged without ever actually existing.

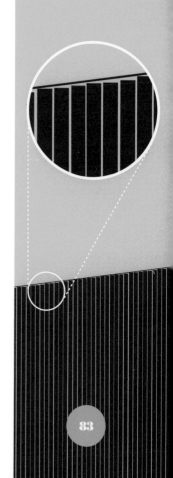

This was the version of infinity that powered calculus, which involves manipulating an infinite set of infinitely small entities. But at the end of the nineteenth century, German mathematician Georg Cantor combined the ideas that Galileo had first considered with set theory, to examine some of the properties of "real" infinity, including the discovery that one infinity can be bigger than another. In this chapter, the topics go large.

TIMELINE

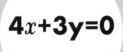

$$4x+3y=0$$

AL-KHWARIZMI
The Persian philosopher Al-Khwarizmi writes the *Hisab al-jabr w'al-muqabala* (*Compendious Book on Calculation by Completion and Balancing*), which will be translated into Latin as *Liber Algebræ et Almucabola* by the English scholar Robert of Chester in 1145.

NEWTON AND CALCULUS
Mathematician and natural philosopher Isaac Newton is forced to spend two years at home in Lincolnshire when Cambridge University is shut down due to the plague. During this time, Newton develops his first ideas on calculus.

c. 240 –270

c. 820

1557

1665

EARLY ALGEBRA
Greek philosopher Diophantus produces 13 books of *Arithmetica*, containing a series of algebraic problems. The six surviving books will help shape the development of algebra in Europe—and it is in the margin of a copy of *Arithmetica* that Fermat will write his "last theorem" claim.

PLUS AND MINUS
Welsh mathematician Robert Recorde introduces the = sign saying "I will sette a paire of paralleles . . . because no 2 things can be moare equalle." He also introduced the + and – signs to Great Britain, though they were already in use in Germany.

INFINITY

Bohemian priest and philosopher Bernard Bolzano's work on infinity, including infinite sets and series, is published posthumously. In his book *Paradoxien des Unendlichen* (Paradoxes of the Infinite), Bolzano builds on Galileo's early observations, laying the ground for Georg Cantor's groundbreaking work on infinity.

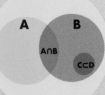

NOETHER'S FIRST THEOREM

German mathematician Emmy Noether proves her "first theorem," which demonstrates the relationship between symmetry and conservation laws. Published in 1918, this will become one of the most important papers in the establishment of particle physics and the unification of physical forces.

1684 **1848** **1874** **1915**

LEIBNIZ VS. NEWTON

German mathematician Gottfried Wilhelm von Leibniz publishes his own version of calculus after meeting with Isaac Newton. Despite little evidence, 24 years later Leibniz is accused by the Royal Society of plagiarism. Leibniz's notation becomes the standard, as does his name for it—calculus.

SET THEORY

German mathematician Georg Cantor publishes the first of a series of papers that establish set theory, which will be used both as the basis for his later work on infinity and as the most basic building blocks for number-based mathematics.

BIOGRAPHIES

AL-KHWARIZMI (c. 780–c. 850)

There are few certain biographical details of Muhammad ibn Musa al-Khwarizmi who was possibly born in Baghdad. He was a scholar in the House of Wisdom there, an academy established by the caliph Al-Mamun, to whom Al-Khwarizmi's manuscript on algebra *Hisab al-jabr w'al-muqabala* (from which the word "algebra" originates) was dedicated. His interest in algebra was at least in part driven by the complexity of Islamic inheritance laws; Al-Khwarizmi stated that its applications included "inheritance, legacies, partition, lawsuits, and trade." The two terms in the title refer to simplifying equations by combining identical powers (*al-jabr*, or "completion") and by combining numbers (*al-muqabala*, or "balancing"). The book used both algebraic and geometric methods, perhaps inspired by Euclid's *Elements*. Other influences were Hebrew and Indian mathematical writing. Al-Khwarizmi also wrote on astronomy and geography. However, his other best-known book was translated as *Algoritmi de Numero Indorum* in Latin. Although it was Fibonacci's *Liber Abaci* that established "Arabic" numerals in Europe (see pages 38–39), Al-Khwarizmi's book also explained them, and the Latinized version of his name gave us the word "algorithm."

ISAAC NEWTON (1643–1727*)

Now regarded as a physicist, in his time Isaac Newton was considered primarily a mathematician. Born in Lincolnshire, he went to Cambridge in 1661, but soon after he graduated, the university closed due to an outbreak of the plague. During time spent at home, Newton claimed he came up with many of the ideas that made him famous, including calculus. After returning to Cambridge, he became Lucasian Professor of Mathematics in 1669 (as did Stephen Hawking in 1979). Two years later he was elected a Fellow of the Royal Society. After falling out with the Society's demonstrator, Robert Hooke, Newton withdrew from the scientific mainstream until the 1680s, when astronomer Edmond Halley persuaded him to consider the motion of the planets. Later, Halley paid to have Newton's mathematical masterpiece on motion and gravity published—*Philosophiae Naturalis Principia Mathematica*. In 1696, Newton became Warden of the Royal Mint and engaged little further in science, apart from publishing *Opticks* in 1704, based on work from decades earlier.

* Newton's birthdate is often given as Christmas Day 1642, and his death year as 1726, but that is old-style dating.

EMMY NOETHER (1882–1935)

Although born Amalie, in Bavaria, Germany, Emmy Noether always preferred Emmy. Unusually for someone who became a leading mathematician, she showed no interest in math at school, and initially trained to teach languages. But by 1903 she was studying mathematics at the University of Erlangen and was awarded a PhD in 1907. By 1915, the German mathematician David Hilbert had put her forward for habilitation, the German academic requirement for professorship, barred to women—Hilbert petitioned the government to make a special exception. The work that put Noether among the mathematical greats was her theorem linking symmetry and the conservation laws of physics. She proved that each of these laws, such as conservation of energy, could be derived directly from assuming a symmetry of nature—if, for example, a system continued to act exactly the same way when rotated, then angular momentum, the "oomph" of rotation, was conserved. Noether's theorem would become absolutely central to developments in twentieth-century physics. In 1933, Noether's Jewish ancestry and support for communism led to her losing her position under the Nazi regime. She moved to America, where she died two years later.

GEORG CANTOR (1845–1918)

Born in St. Petersburg, Russia, of a Swedish father and a Russian mother, Georg Cantor moved with his family to Germany when he was 11. In 1862, he started a course at the ETH in Zurich, but when his father died a year later, he moved to the University of Berlin. Two years after receiving his doctorate in 1867, he took a position at the University of Halle. At the time, Halle was famous for music—but not for math. Undoubtedly Cantor saw this as a stepping stone to return in triumph to Berlin, or another university highly regarded for its mathematics, but the response to his ideas would prevent this happening. Cantor's first major discovery was that it was possible to pair off the infinite points on a line with those on a plane, or multidimensional space—something he himself struggled to accept. He went on to develop the hugely influential mathematics of set theory and, expanding on his work on infinite sets of points, the concept of transfinite numbers. His work was considered outrageous by some, notably the German mathematician Leopold Kronecker who went out of his way to damage Cantor's career. Cantor suffered from depression and in later years spent increasing amounts of time being treated in sanatoria.

EQUATIONS

THE MAIN CONCEPT | If numbers and arithmetic got mathematics started, equations made it soar. The principle is simple—an equation tells us that two things are equal. Typically, an equation will be algebraic, having a combination of variable and fixed quantities, though this isn't necessary. Equations are often compared to old-fashioned weighing scales that work as a balance. When the balance is level, we know that the weight on one side equals the weight on the other, even though, say, one side contains apples and the other, metal weights. Equations can also appear in geometry, as became apparent once Descartes had brought geometry and algebra together with Cartesian coordinates (see page 66). Many shapes can be represented by equations—a simple circle, for example, can be represented as $x^2 + y^2 = r^2$, where r is the radius. Similarly, an equation can feature trigonometric functions, such as $x = sin(y)$. Equations are solved when conditions force one or more of the variables (x and y in the examples above) to have particular values. Depending on the application, solving equations can be used to do anything, from working out the interest in a bank account to the path of a spaceship. Often this involves "simultaneous" equations, where two or more equations are combined to fix the values of the variables (see opposite).

DRILL DOWN | Simultaneous equations bring together multiple equations to provide a solution, using one equation per unknown variable. If we have a single variable, one equation suffices. It's easy to work out x in $2x = 4$ by dividing both sides of the equation by 2. But $2x + y = 5$, has an infinite set of values for x and y. For example, x could be 2 if y was 1, or x could be 1 if y was 3. However, if we also know that $x + y = 3$, then we can replace y in the first equation with $3 - x$ and the only possible values for x and y are 2 and 1 respectively.

BASIC ARITHMETIC; NEGATIVE NUMBERS
Page 26

CARTESIAN COORDINATES
Page 66

BASIC ALGEBRA
Page 90

FOCUS | *The ancient Greeks focused their mathematical expertise on the visual and spatial in geometry. It's not surprising that they didn't develop the concept of equations—not only did they lack the symbols to produce something like $A + B = C + D$, they would have written them without any spacing. Their equivalent would be* theaandthebtakentogetherare equaltothecanddtakentogether.

BASIC ALGEBRA

THE MAIN CONCEPT | Basic algebra consists of simple algebraic equations, such as $3x + 4 = 0$, or quadratic equations, such as $3x^2 + 8x + 4 = 0$, as well as equations containing multiple variables, such as $4x + 3y = 0$ or $4xy = 0$. Algebraic equations are used to work out the outcome of a process represented by the equation by finding solutions for the variables. For equations with a single variable, the number of solutions depends on the "order" of the equation—the highest exponent of the variable indicates the maximum number of solutions. So $3x + 4 = 0$ is order 1 (as $3x$ is the same as $3x^1$) and has a single solution, easily worked out by taking 4 from both sides of the equals sign, giving $3x = -4$, then dividing both sides by 3 to give $x = -4/3$. Quadratic equations, such as $3x^2 + 8x + 4 = 0$, have two solutions. These can often be worked out as a pair of multiplied mini-equations—in this case $(3x + 2)(x + 2) = 0$. If we give x the value $-2/3$ in the first bracket or -2 in the second it makes the value 0, so these are the solutions of the equation. Usually, a formula is used to work out the solution.

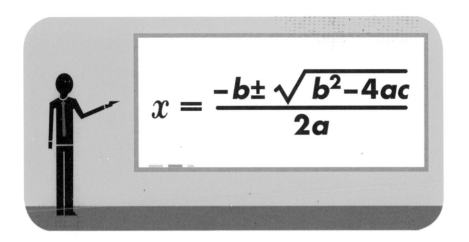

$$x = \frac{-b \pm \sqrt{b^2 - 4ac}}{2a}$$

DRILL DOWN | Although the quadratic equation features heavily in school math it has relatively few applications—it's best seen as training wheels to become familiar with algebra. Even so, when in 2003, the UK's National Union of Teachers suggested teaching the equation was cruel to schoolchildren, mathematicians leapt to its defense. They pointed out that it had been used to calculate crop yields, and that the irrational square root of 2 (which is, for example, the ratio of the sides in the European A paper sizes) is the solution to the simple quadratic equation $x^2 = 2$. On a larger scale, the inverse square laws governing both gravity and electromagnetism are governed by quadratic equations.

FOCUS | *Although problems that involved quadratic equations, such as working out the lengths of the sides of a rectangle, given its area and perimeter, date back around 4,000 years, the first explicit formula for one solution to a quadratic equation was given by Indian mathematician Brahmagupta in 628 CE, while Persian mathematician Al-Khwarizmi correctly identified the familiar formula for both possible solutions in the ninth century.*

POWERS, ROOTS & LOGARITHMS
Page 34
EQUATIONS
Page 88
LINEAR ALGEBRA
Page 92

LINEAR ALGEBRA

THE MAIN CONCEPT | Linear algebra sounds as if it should be the simplest and least useful form of algebra, as it involves only adding component parts together with no powers involved. At its simplest, this could be something like $3x = 6$, though a more likely equation would be $3x + 4y - 7z = 0$. In practice, though, linear algebra underlies many important mathematical methods, including the manipulation of "vector spaces"—collections of mathematical quantities that have both size and direction—and the mathematics of matrices, which are rectangular arrays of numbers that are treated as entities in their own right and have their own system of arithmetic. Both vectors and matrices are widely used in modern physics. For example, vectors are used in field theory and to represent forces, while matrices are important in both the symmetries underlying particle physics and in quantum theory, particularly in components known as eigenvalues and eigenvectors. Eigenvectors are special single-column matrices that transform a matrix in a linear fashion, producing a result that is a multiple of the eigenvector: that multiplier is the eigenvalue. And in the modern approach to geometry, objects involving straight lines, such as lines, planes, and rotations, are defined in terms of linear algebra.

DRILL DOWN | Eigenvalues and eigenvectors are common in quantum physics, but also turn up in Google's PageRank algorithm, used to set the order of its search results. PageRank (allegedly named after Google founder Larry Page) weights pages depending on how many pages point to them and how well ranked those pages are in turn. For example, if CNN's trusted website links to your site, that link is more valued than a link from a blog. But to get CNN's rank, we have to already have ranked every website pointing to it. The algorithm sets up a matrix of rankings: a complex task. The eigenvector of the matrix with eigenvalue 1 turns out to be a score for the page rank of each page.

EQUATIONS
Page 88
VECTOR ALGEBRA & CALCULUS
Page 104
MATRIX MANIPULATION
Page 134

FOCUS | Eigen *is a German word meaning "proper," "peculiar," or "characteristic." Its use in the strange cross-language words "eigenvalue" and "eigenvector" is probably the closest mathematics and physics come to the way that "-gate" is now applied following Watergate. As well as these two there are also eigenfunctions, eigenloads, eigenperiods, eigensolutions, eigenstates, eigentones, and eigenvibrations.*

INFINITY

THE MAIN CONCEPT | Infinity can be regarded in two ways: as a limit that is never reached, or as the size of an infinite set—a set with no last member. Aristotle argued that infinity was like the Olympic Games. The Olympic Games certainly exist—but unless they are on here and now, I can't show them to you; they are a potential thing. Aristotle argued that infinity was similarly a potential entity. It exists. For example, there's no end to the integers. If there were a biggest integer, then we could simply add 1 to it to get an even bigger one. Yet we can't directly observe infinity. It was this potential infinity that formed the limit in calculus and that is represented by the leminscate ∞. However, in his physics masterpiece, *Two New Sciences* (1638), Galileo spends a little time thinking about infinity, and he describes the real thing. Galileo points out some of the strange behavior of infinity. Add 1 to infinity—and you get the same value. Double infinity … and it's still infinity. Galileo also pointed out that you can have infinite collections of things that seemed simultaneously the same size and different sizes. For example, the infinite set of squares is the same size as the infinite set of integers—every single integer has a corresponding square—even though there are plenty of integers that aren't squares, such as 2, 3, 5, 6, 7, 8, 10, and so on.

DRILL DOWN | The strange arithmetical behavior of infinity is illustrated by "Hilbert's Hotel," named after German mathematician David Hilbert. Imagine a hotel with an infinite set of rooms. (Note that though an infinite set is fine, we shouldn't refer to an infinite number of rooms: infinity is not a number.) The hotel is full. But this isn't a problem. Just move the person in room 1 to room 2. The person in room 2 goes to room 3, and so on throughout the hotel. Everyone is accommodated and there's a room spare. Then an infinite coachload turns up. No problem. Just move everyone into even numbered rooms. The infinite set of odd numbered rooms are now free.

FOCUS | *German mathematician Georg Cantor, who both developed set theory and explored Galileo's "real" infinity, introduced two new symbols for true infinity to distinguish it from potential infinity. An aleph (ℵ), the first letter of the Hebrew alphabet, represents the cardinal (counting) infinity, while the small omega (ω) represents the ordinal (ordering) infinity—the limit of the series first, second, third . . .*

SERIES
Page 32
SET THEORY
Page 106
TRANSFINITE VALUES
Page 108

FACTORIALS & PERMUTATIONS

THE MAIN CONCEPT | Much algebraic work makes use of simple arithmetic operations, but some extras have proved particularly valuable. The factorial is a shorthand for a particular repeated operation. Designated using an exclamation mark, it involves multiplying a positive integer by each smaller integer until 1 is reached. (By convention, 0! is 1.) As mathematical symbols go, this is a modern one, introduced in 1808. So, for example, 5! (pronounced "five factorial") is 5 × 4 × 3 × 2 × 1 = 120. Factorials rapidly explode in size. By the time we get to 10! the value is already 3,628,800. The most versatile application of the factorial is in permutations—the different ways that a set of objects can be arranged. For example, if we have three objects, A, B, and C, they can be arranged as: ABC, ACB, BAC, BCA, CAB, and CBA. That's 6—or 3! ways. Similarly, A, B, C, and D can be arranged 4!—24—ways, and so on. Factorials also often crop up as a divisor when calculating combinations—which are subsets of items selected from a longer list, in which we don't care about the order of the items (a "combination lock" should be called a "permutation lock"). There are non-integer equivalents to factorials, but they require calculus to function and are rarely used.

FOCUS | *The exclamation mark (known to programmers as a "shriek") was not universally popular. English mathematician Augustus De Morgan complained: "Among the worst of barbarisms is that of introducing symbols which are quite new in mathematical, but perfectly understood in common, language . . . the abbreviation n! . . . gives their pages the appearance of expressing admiration that 2, 3, 4, etc. should be found in mathematical results."*

DRILL DOWN | We often want to find how many combinations of, say, three items it's possible to make from a longer list, but don't care about their order. If we've got L items and want to pick s of them, we ignore the rest by using $L! / (L-s)!$—but this counts each possible order of s items. To ignore the duplicates, we divide by s! When picking 3 items from a list of 6, the total options are $6!/3! = 120$, and ignoring order we divide by 3! to get 20. This is written mathematically as:

$$\binom{n}{k} = \binom{6}{3} = \frac{n!}{k!(n-k)!} = \frac{6!}{3!(6-3)!} = 20$$

If you set n to 59 and k to 6, the answer is 45,057,474—the chance of winning the jackpot in a lottery with 6 numbers drawn from 59.

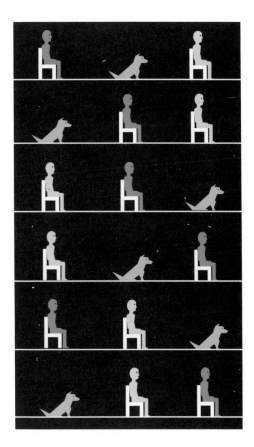

DIFFERENTIAL CALCULUS

THE MAIN CONCEPT | Isaac Newton's masterpiece, *The Principia*, which gives us his three laws of motion and his work on gravitation, is full of geometry—but to get his results Newton used a new approach he called the "method of fluxions," now called calculus, the name given to it by Newton's competitor, Gottfried Leibniz. The "differential" part distinguishes it from integral calculus (see page 100): the two approaches to calculus are inverses of each other. Differential calculus is used to work out the rate of change of one variable with respect to another. For example, imagine you are trying to work out the acceleration of a car. If the car's speed increases steadily, a plot of the speed against time is a straight line, and the acceleration is just the slope of that line—the change in speed divided by the change in time. But what if, for example, the speed increases with the square of the time? Now, the plot is curved. But imagine zooming in to a very tiny part of the curve—it's almost a straight line, and you can use the slope as before. Differential calculus takes smaller and small segments which, at the "limit," are infinitesimally small; at that point the actual value drops out. This is extremely powerful whenever one value (here, speed)—varies with another (here, time).

DRILL DOWN | In his method of fluxions, Newton used "pricked notation," where the rate of change was represented by a dot over the symbol for the thing changing. This was difficult to read and many equations in calculus don't involve time. In Leibniz's notation, still used today, the way x changes with time is shown as dx/dt, while the way that x varies with y is dx/dy. Both methods involved having a small change (represented as δx by Leibniz and o by Newton) that disappears, and could be worked using simple rules. So, for example, if $y = 2x^3$, differentiating it we get $dy/dx = 6x^2$—multiplying by the exponent and reducing the exponent by 1.

FOCUS | *When one of Newton's associates, John Keill, published a paper accusing Gottfried Leibniz, the co-discoverer of calculus, of plagiarism, Leibniz was dismayed and asked the Royal Society to establish the truth. The Royal Society set up an 11-man committee, which came down on Newton's side. Not surprising, as the report's author was the President of the Royal Society—Isaac Newton.*

INTEGRAL CALCULUS

THE MAIN CONCEPT | Where differential calculus uses disappearingly tiny changes to see how one variable changes with respect to another, integral calculus divides a shape up into disappearingly narrow segments to calculate the overall area (or equivalent in more dimensions). So, for example, to find the area of a circle, it can be divided up into slices. If more and more slices are used, the curves at the top and bottom of the slice become closer and closer to straight lines: in the "limit" of taking an infinite set of infinitely thin slices, the exact value is obtained. Approximating this way with narrow rectangles was used in ancient times before the idea of taking it to infinitesimal limits was introduced. Integral calculus is the inverse of differential calculus. As this "integration" involves adding up slices, it is represented by an S for "sum"—but that S is stretched to form a kind of bracket $\int$, which is usually annotated with small numbers top and bottom showing the range over which the variable value is being considered. Integrals are used both in finding areas and volumes, but also widely in physics to reverse the kind of operation resulting from differential calculus—so, for example, given the varying velocity of a moving body, integration will give you how far it has traveled.

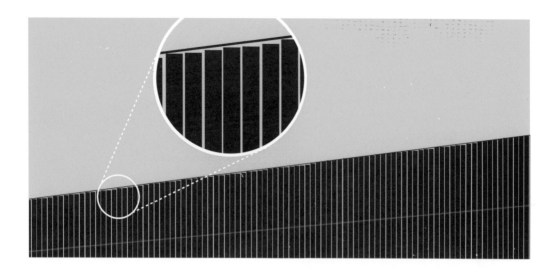

DRILL DOWN | When Newton and Leibniz first introduced calculus, the philosopher Bishop George Berkeley pointed out that there were serious problems with the approach as it often involved dividing a value by another value, each of which were allowed to become zero—and zero divided by zero is meaningless. At the time, this was mostly swept under the carpet, as calculus worked and was extremely useful. Mathematicians Augustin-Louis Cauchy and Karl Weierstrass would each refine the approach until the problem was removed because instead of calculating a final result it was seen as the limit of a process which made the disappearing quantity closer and closer to zero without ever actually reaching it.

CONIC SECTIONS
Page 60
DIFFERENTIAL CALCULUS
Page 98
VECTOR ALGEBRA & CALCULUS
Page 104

FOCUS | *Integrating often involves a range that runs to infinity, which can result in interesting results. The shape made by plotting a graph of 1/x for every x bigger than 1, rotated around the axis to make a long, pointed object known as Gabriel's Horn, has the odd property of having a finite volume, but an infinite surface area.*

FUNCTIONS

THE MAIN CONCEPT | Given its modern notation by Swiss mathematician Leonhard Euler, a function is a powerful mathematical concept that applies a "black box" operation to a number or more complex mathematical structure. A function involving the variable x is written $f(x)$—pronounced "f of x." The function's internal workings can contain any mathematical structure that uses x. That might be very simple—for example, it could be $2x$. In that case, whatever value x has, $f(x)$ has twice that value. But it could also be a 1,000-line-long piece of mathematics. We replace that mathematics with $f(x)$, and to get the result, we plug the desired value for x into that mathematical structure. Functions are like maps linking inputs (the domain) and outputs (the range). Functions occur all the time in mathematics and physics and have analogs in the physical world. So, for example, on a toaster with a dial that has positions corresponding to different amount of browning of toast, the degree of browning is a function of the position of the dial. Functions are also the main structural unit of computer programming languages. Closely related to functions are operators. Effectively, an operator is a mechanism for applying something like a function to a whole collection of variables or places in a space or field at once.

DRILL DOWN | Computer programs are usually written in a programming language. This is an intermediary between the human programmer and the manipulation of 0s and 1s that goes on within the computer's processor. It was soon realized that it would be useful to break up the code for a program into separate chunks that handled specific aspects of the job as a whole. These could then be used many times within the program and also had the added benefit of being reusable in other programs. Most such subroutines are functions, acting like a mathematical function. They produce a result and have settings determined by variables that are fed into the function, just as the x is fed into $f(x)$.

FOCUS | *The word "function" is one that non-mathematicians think of as "what something does." Mathematics makes use of a very specific aspect of this definition— executing a task. It was first used mathematically by Gottfried Leibniz to describe the relationship of a straight line to a curve (e.g., tangents), but the mathematician Johann Bernoulli expanded its use to its present form.*

VECTOR ALGEBRA & CALCULUS

THE MAIN CONCEPT | Standard algebra and calculus deal with "scalar" values—just numbers—and the way they change. However, many natural quantities are vectors—they have both size and direction. It was realized in the nineteenth century that the mathematics that worked so well with scalars could be extended to vectors. Basic vector algebra is linked to linear algebra, while vector calculus typically applies to changes in a field, each point of which has a vector value. The simplest aspect of vector algebra is addition—as vectors have both size and direction, this addition is visualized by placing two arrows on a chart, the second starting at the point of the first—the result, produced by drawing an arrow from the tail of the first arrow to the point of the second, is the result of the sum. Vector calculus has three main operations: grad (gradient), div (divergence), and curl. The gradient is the equivalent of differentiation, but applied to every point in a field of scalar values simultaneously. Divergence produces a scalar field based on the source of each point in a vector field, while curl measures the amount of rotation at each point in a vector field. Vector calculus is essential for calculating everything from heat flow to the effects of electromagnetism.

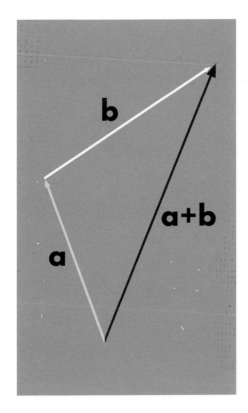

DRILL DOWN | The symbol used for the vector calculus operator ∇ is called "del"—it's an upside down Greek capital delta. (The symbol is also sometimes called "nabla," as it resembles the shape of an ancient Greek harp of that name.) The basic symbol is used directly to represent grad (∇f); with a dot to indicate the process is a "dot product" producing a scalar result for div ($\nabla \cdot F$); and with a cross, called a "cross product," producing a vector result as curl ($\nabla \times F$). Just as standard differentiation has an inverse in integration, each of the vector calculus operations has an integrating inverse.

FOCUS | *The terms gradient, divergence (or, to be precise, its opposite, convergence), and curl were devised by James Clerk Maxwell while he was working on a predecessor to vector calculus called "quaternions." Maxwell originally wanted to call curl "twirl" (also considering whirl, twist, and turn) but settled on curl because he thought mathematicians might consider twirl "too dynamic."*

SET THEORY

THE MAIN CONCEPT | Despite being developed much later than arithmetic, set theory is the fundamental theory on which mathematics is built. A set is a collection of items known as members or elements—they can be anything from physical objects (the set of all elephants, for example), to sets in their own right. The integers are defined using sets. Zero is an empty set—a set with nothing at all in it—represented by Ø; 1 is a set containing the empty set: {Ø}; 2 is a set containing the empty set and the set containing the empty set: {Ø, {Ø} } . . . and so on. A term from set theory has escaped into common usage: the "subset." A subset consists of some, but not all, of the elements of a larger set. So, for example, the even numbers form a subset of the integers. As with mathematical logic, sets and subsets can be represented using Venn diagrams, where ∩ indicates an intersection, and ⊂ indicates a subset. Although set theory is constructed logically, one of the axioms it is built on, the axiom of choice, is difficult to justify, and the ability to make a set a member of itself causes significant paradoxes. Nonetheless it is generally agreed that set theory provides our best fundamental theory to build mathematics on.

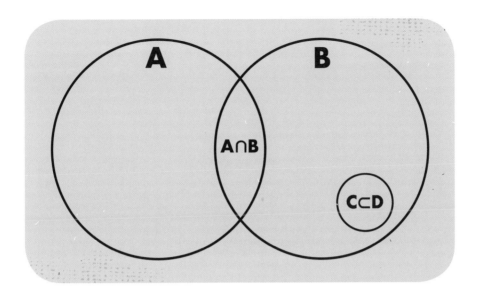

DRILL DOWN | Philosopher Bertrand Russell demonstrated the paradoxical difficulty of set theory by envisaging the set of "all sets that are not members of themselves." Sets can be members of themselves. For example, the set "cars" is not a member of itself—the set is not a car—but the set "things that aren't cars" is a member of itself, as long as we're happy that a set is a thing. The question is whether "all sets that are not members of themselves" is a member of itself. If it is, then it isn't. And if it isn't, then it must be. It's a bit like interpreting "This is a lie."

FOCUS | *The inability to properly fix the problem with the axioms of set theory does not show that set theory is a failure as the foundation of mathematics. The mathematician Kurt Gödel proved in his "incompleteness theorem" that for a non-trivial system of mathematics no consistent system of axioms is capable of proving everything required—there will always be gaps.*

TRANSFINITE VALUES

THE MAIN CONCEPT | It might seem that by definition there can be nothing bigger than infinity, but Georg Cantor showed otherwise. The size of a set is its cardinality, determined by pairing off members of one set with members of another. If we can go through the whole set and do this, with none left over, the sets have the same cardinality. Starting with the infinity of the integers, Cantor showed that the infinity of every rational fraction had the same cardinality. He did this by imagining a table including every rational fraction and finding a route that, step by step, led through the table—one possible route is shown by the arrows in the diagram opposite. Then each item in the table could be paired off with the integers. Cantor called their shared cardinality $\aleph_0$ (aleph null). He then found a way to attempt the same with, for example, all numbers between 0 and 1. If we can produce a list of such numbers, we can pair it off with the integers. Cantor imagined scrambling the list (for easier visualization). He then took the first digit of the first number, the second digit of the second number, and so on, and added 1 to each digit. The new number created this way was not in the list, demonstrating that it's impossible to pair off every number between 0 and 1 with the integers. This "transfinite" cardinality is $\aleph_c$—where "c" stands for continuum.

DRILL DOWN | Cantor struggled for years attempting to determine whether or not $\aleph_c$ was the same as $\aleph_1$—the next infinity up from $\aleph_0$. He showed that $\aleph_c$ was the power set of $\aleph_0$. A power set is the set of all the subsets of a set, which has cardinality 2^n, where n is the cardinality of the set. But Cantor could not prove $\aleph_c$ was the next biggest infinity, something that combined with vehement opposition from mathematician Leopold Kronecker seems to have made worse his mental problems. What Cantor would never know is that the mathematician Kurt Gödel would demonstrate that it was impossible to prove whether or not Cantor's so-called continuum hypothesis was true.

FOCUS | *Ordinal transfinite values have their own, different structure; when considering order, a one-to-one correspondence is harder to establish. Where both {a1, a2, a3 ...} and {a2, a3 ...} have ordinal value ω, {a2, a3 ... a1} has ordinal value ω + 1. Cantor established a whole hierarchy of the ordinals, including ω raised to the ω repeated ω times, which was called ε0, and so on.*

SERIES
Page 32
INFINITY
Page 94
SET THEORY
Page 106

SYMMETRY

THE MAIN CONCEPT | In ordinary English usage, "symmetry" means mirror symmetry, where one side of a view is the mirror image of the other. Letters such as A and T are mirror symmetric, whereas R and G, for example, aren't. But in mathematics, symmetry has a wider definition. Something has symmetry if it stays the same when it undergoes *any* transformation. So, for instance, there is rotational symmetry—where something still looks the same if you rotate it. A circle has full rotational symmetry as it can be rotated by any angle on a plane and it stays the same, whereas a square only has rotational symmetry for rotations of 90 degrees. Another example is translational symmetry, where you shift your viewpoint sideways and the result looks the same. A repeated pattern, for example, has translational symmetry where a random pattern doesn't. In a theorem linked to a field of mathematics known as the "calculus of variations," the German mathematician Emmy Noether showed mathematically that if such symmetries could be applied to a system, then within that system there must be conservation of a physical property. For example, if there is translational symmetry through space, momentum must be conserved, while with translational symmetry through time, energy is conserved.

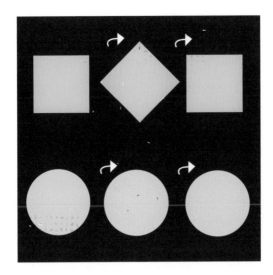

DRILL DOWN | Noether's theorem for symmetry uses the calculus of variations, which is often used to find maximum and minimum values and is important in the principle of least action, which is central to much of physics. Newton's laws of motion, for example, can be formulated from a version of the principle that says a moving object will take the path where the kinetic energy minus the potential energy integrated along the path is minimized. The related "principle of least time" says that light will take the path that minimizes its journey time, which is why it bends when passing, say, from air into water, spending less time in the medium where it is slower.

FOCUS | *Thanks to Noether's work, symmetry and symmetry breaking would become one of the great drivers of twentieth-century physics. It was symmetry considerations, for example, that led to the idea that particles previously considered to be fundamental, such as protons, had subcomponents. Symmetry was also at the heart of the unification of the electromagnetic and weak nuclear forces in so-called electroweak theory.*

PERSPECTIVE & PROJECTION
Page 64
INTEGRAL CALCULUS
Page 100
GROUPS
Page 112

GROUPS

THE MAIN CONCEPT | There's no real distinction in the English language between a set and a group, but mathematicians like to be precise, and they use "group" to mean a particular type of set—a set where there is an operation that will allow you to take any two members of the group and produce a third (with a few other technical restrictions). The integers are a group because using an addition operation on two integers will always produce an integer. Another simple group is any set of integers used in modular arithmetic. Groups form a bridge between the concepts of sets and symmetry. When an object is symmetrical (using the mathematical definition), there will be a group called a "symmetry group," which is a group of matrices that define the different ways the object can be transformed without changing it. There is a standard notation describing the different symmetry groups. So, for example, the symmetry group for the rotations of a sphere is called SU(3). Symmetry groups are particularly important in physics. For example, one type of symmetry group, called a Lie group, which deals with continuous symmetries (like that of a sphere, rather than, say, a mirror symmetry) is widely used in particle physics.

DRILL DOWN | The concept of quarks, the fundamental particles making up protons, neutrons, and particles called "mesons," emerged from applying a symmetry group. Physicists had identified a symmetry in particles which were ascribed to two properties known as strangeness and isospin. The particles were distributed on diagrams in groups of eight, which the originator of the quark concept, Murray Gell-Mann, referred to as the "eightfold way" in a reference to a Buddhist concept. Mathematically, these patterns were suggestive of the symmetry group SU(3). Gell-Mann suggested that the group comprised three "flavors" of underlying particles which he called quarks (pronounced "kworks")—up, down, and strange. A second generation of quarks—charm, top, and bottom were added later.

FOCUS | *An early developer of group theory was the tragic French mathematician Évariste Galois, who gave groups their name and devised the Galois group, which links groups and fields. He no doubt would have done much more, but he died aged only 20 as a result of a duel, probably on behalf of a young woman called Stéphanie-Félicie Poterin du Motel.*

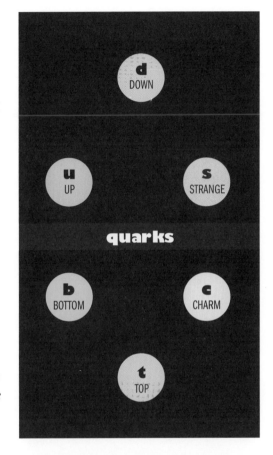

"If you want to be a physicist, you must do three things—first, study mathematics, second, study more mathematics, and third, do the same."

ARNOLD SOMMERFELD IN AN INTERVIEW WITH
PAUL KIRKPATRICK, QUOTED IN DANIEL KEVLES,
THE PHYSICISTS (1978)

4

APPLIED MATHEMATICS

INTRODUCTION

It is perfectly possible to construct a system of mathematics that bears no resemblance to the real world. Although the origins of math may have been as a reflection of familiar objects, mathematics has long since become detached from the physical realm. As long as a mathematical system is self-consistent, not breaking its own rules, you can do what you like. If you want to make $2 + 2 = 5$, that's fine, provided this does not conflict with the rest of your mathematics.

An unreal universe

We have already seen a number of aspects of unreal math. There is no square root of a negative number in the physical world—even a negative number requires a degree of interpretation. There are only four physical dimensions, yet mathematicians are happy to work in thousands. It's debatable whether infinity has any real-world equivalent. So, many mathematicians spend their time working in what appears to be an isolated mathematical universe with no consequence other than its intellectual challenge. As Wigner observed, "Most more advanced mathematical concepts ... were so devised that they are apt subjects on which the mathematician can demonstrate his ingenuity and sense of formal beauty." And yet, as he also said, mathematics is "unreasonably effective in the natural sciences." For something mostly built on highly theoretical foundations, it has proved remarkably useful.

It should be no surprise that the late Stephen Hawking worked not in the physics laboratory at Cambridge but in the Department of Applied Mathematics and Theoretical Physics. Modern physics is driven by math to the extent that it seems many physicists are indistinguishable from pure mathematicians, dreaming up toy universes of black hole firewalls and superstrings for which there is no evidence as yet from science, only mathematical challenges. Similarly, the information and computing architecture that underlies our modern world has math at its heart.

Math for the real world

In this book, physics and computer science are often used as examples of the use of math, but as we will see in this chapter, there are also more mundane ways that math makes its presence felt in everyday life. Whether it's the bookie calculating odds for the next horse race, an actuary setting insurance premiums (or, these days, providing an input to the algorithms that will do so), or a weather forecaster putting together tomorrow's outlook, math is central to the job.

Even the "soft" sciences such as economics, sociology, and psychology make regular use of mathematics to model the economy or reflect human behavior. This was not really possible before the emergence of probability and statistics, which as well as helping bookies and insurance workers has been central to applying numbers to the general field of human activity. Less well known is the field of operations research, developed for military purposes during World War II and expanded since to do everything from working out the best queuing systems to scheduling airlines.

All these applications would originally have made use of the mathematician's traditional tools of paper and pencil. However, since the end of the war, applied mathematics has increasingly involved computers to perform analyses and calculations that would be beyond manipulation by hand. Famously, Charles Babbage, the inventor of early mechanical computers, was said to have been inspired by helping a friend, John Herschel, to laboriously construct tables of numbers. Babbage is said to have cried out, "My God, Herschel, how I wish these calculations could be executed by steam!"

It's worth noting, though, that computers are not just the main work tools of applied mathematicians—they are themselves entirely dependent on mathematics. An electronic computer's universe is a sea of 0s and 1s, undergoing constant mathematical processes. The computer (or the smartphone in your pocket) is the embodiment of applied math, transformed from manipulation of numbers and logic into the bewildering array of applications that modern technology delivers.

In this chapter we will see math that encroaches on human behavior and the ways in which pure mathematicians' wild flights of fancy have been tamed, but also the situations where raw mathematical power seems able to overcome the ordered world we expect, whether it is in the unmasked randomness of probability-driven quantum theory or the wild power of chaos, unleashed in the theory that explains the behavior of everything from the stock market to the weather. To a pure mathematician, applied mathematics may seem a touch grubby and mundane, but to the rest of us it is where math gets its hands dirty and gets the job done.

TIMELINE

FOURIER SERIES
French mathematician Jean-Baptiste Joseph Fourier introduces what will become known as Fourier series in his *Théorie Analytique de la Chaleur* ("The Analytical Theory of Heat"), showing that complex and discontinuous waveforms can be broken down to constituent simple waves in a so-called Fourier transform.

THE MATRIX
English mathematician James Joseph Sylvester coins the term "matrix" (Latin for "womb") for a two-dimensional array of numbers. Such arrays had been in use for around 200 years, but become a major tool of mathematics around this time.

1663 — **1822** — **1822** — **1850**

PROBABILITY
Giralamo Cardano's book *Liber de Ludo Aleae* ("Book on Games of Chance") is published around 100 years after it was written and 87 years after his death. It is the first proper analysis of probability, but is initially ignored because it deals with betting odds and cheating.

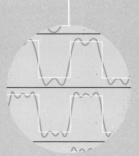

EARLY COMPUTERS
British inventor Charles Babbage completes a model of his Difference Engine, a mechanical calculator for the production of astronomical tables. He is funded to build a complete engine, but engineering limits and distraction by designing his (never begun) programmable computer, the Analytical Engine, mean that he never finishes.

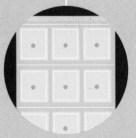

OPERATIONS RESEARCH

British physicist and radar pioneer Albert Rowe, working on radar development at the Bawdsey Research Station (Suffolk, England), introduces the formal concept of operations research (OR), although basic OR principles date back to the 1840s in work on mail sorting and the safety of railway wagons.

CHAOS THEORY

American mathematician Edward Norton Lorenz is inspired to develop chaos theory when he inputs the starting values for a weather-forecasting model on his LGP-30 computer and finds a small rounding error produces a totally different forecast.

1928   **1937**   **1944**   **c. 1960**

GAME THEORY

Hungarian-American mathematician John von Neumann publishes his paper *On the Theory of Parlor Games*, which begins the field of game theory, describing human interactions based on strategies for playing simple games. These will be employed in the development of the nuclear "mutual assured destruction" strategy.

CODE-BREAKING

Colossus, the world's first programmable electronic digital computer, goes into operation at the British Bletchley Park code-breaking center. The device and its successors are destroyed after the war on Churchill's orders, and their pioneering work not revealed until the 1970s.

BIOGRAPHIES

JACOB BERNOULLI (1655–1705)

The oldest of the great Bernoulli mathematical dynasty, also including two Nicolauses, three Johanns, and a Daniel, Jacob (also known as Jacques) was born in the Swiss town of Basel. He studied theology at the University of Basel and was ordained, but by his early twenties he was convinced that mathematics was the most important thing in his life. After several years touring Europe, he returned to Basel where he taught mechanics and began to work in mathematics. By 1687, Jacob had become a professor of mathematics. He worked together with his brother Johann for a number of years, but after a decade their mutual support had turned to bitter rivalry. Like a number of his relations, Jacob made significant contributions to probability theory, most notably in his book *Ars Conjectandi*, published eight years after his death. The book established much of the basics on permutations and combinations, as well as significant aspects of probability, such as an early version of the "law of large numbers," which says that the more times an experiment is run (provided it is fair), the closer the average outcome will be to the expected value.

ABRAHAM DE MOIVRE (1667–1754)

Born in France, Abraham de Moivre first studied logic at college in Saumur, which gave him an interest in mathematics, after which he moved to Paris to study physics. This was a time of increasing persecution of Protestants in France—by the time he was 20, de Moivre had moved with his brother to London. Here, while employed as a private math tutor to the sons of wealthy families, he came across Newton's recent *Principia* and would come to know both Newton and his supporter Edmond Halley. After some early work extending some of the math in the *Principia*, de Moivre became a Fellow of the Royal Society, but he remained an under-funded private tutor throughout his career. His biggest contributions to mathematics were in the new field of probability theory. He was instrumental in the introduction of two of the most commonly used distributions in probability—the normal distribution and the Poisson distribution (which he worked on before Siméon Denis Poisson). These explain the way randomly varying characteristics are distributed for continuous options or those coming in integer chunks respectively, helping do everything from check errors to forecast behavior.

PATRICK MAYNARD STUART BLACKETT (1897–1974)

Born in London, Blackett was technically a physicist, doing significant work on cloud chamber research, but his biggest contribution was in applied mathematics. While serving in the British navy in World War I, Blackett noticed the poor use of artillery and was inspired to take an interest in physics and math. Given an opportunity to study at Cambridge, he became an experimental physicist, working with Ernest Rutherford. From there he went on to Birkbeck College, London, and Manchester University in physics professor roles. However, he did not forget his interest in mathematics and by 1935 had joined the Aeronautical Research Committee, steering the development of radar. He worked at the Royal Aircraft Establishment in Farnborough before helping develop a new field of applied mathematics known as operations research. This involved using mathematical methods to improve the effectiveness of military operations, including the best way to structure convoys to minimize the risk of sinking and devising the optimal distribution of armor plating for aircraft. Blackett was elected to the Royal Society and became a life peer in 1969.

JOHN VON NEUMANN (1903–1957)

Born in Budapest, Hungary, as János, John von Neumann was a leading figure in the Manhattan Project to develop the atomic bomb, was central to the US development of electronic computers, and worked widely in mathematics, from topology to game theory. Von Neumann was a child prodigy, already using calculus at the age of 8; by 23 he had simultaneously achieved a doctorate in mathematics from Pázmány Péter University in Budapest and a degree in chemical engineering (a fallback subject at his father's instance) from the ETH in Zurich. He was soon lecturing in mathematics in Berlin, but by 1929 had moved to Princeton in the US, the university that was his home for the rest of his career. Mathematically, Von Neumann was influential in set theory and in the use of operators in quantum mechanics. Just one of his many achievements was the establishment of game theory as a part of mathematics. This finds strategies in games where typically two players can each make a choice that influences the outcome of the game—it would be used extensively to model the interaction between the US and the USSR during the Cold War.

ENCRYPTION

THE MAIN CONCEPT | Keeping sensitive information secure has been a concern as long as there has been writing. This can be achieved by physical concealment, codes (which replace a whole message with words or symbols; FISH could mean "Attack tomorrow at 9am"), and ciphers (which use mathematical manipulation to modify a message). For a long time, such encryption was the business of spies, but now, whenever a secure link is established on the Internet, encryption comes into play. If a website uses HTTPS: (indicated by a padlock symbol in the browser), or we use a secure messaging service, the software employs a mathematical process to make the information unreadable to an eavesdropper. Encryption usually relies on a key. At its simplest, this is a word or phrase, the letter values of which (position in the alphabet) are added to the message. So, for example, HELLO encrypted using the key APPLE adds A to H (1 to 8) giving I (9), P to E, and so on, producing IUBXT. Internet encryption, such as RSA, has dual keys—a public one that anyone can use to encrypt information and a private one that only the owner can use. The result is like a mailbox anyone can post into, but only the owner can open.

DRILL DOWN | It's easy for a computer to multiply huge prime numbers together (say with 32 digits each), but hard to work out the original numbers from the result. In a public key/private key encryption system, the multiplied number is used as a modulus in mathematical manipulations producing the keys. The public key is made freely available; data is encrypted by raising a numerical version of the data to the power of the public key and the result given with a modulus of the multiplied primes. To decrypt, a similar process is used, but raising the encrypted message to the power of the private key. Systems such as Whatsapp employ several public and private keys on each message.

BASIC ARITHMETIC;
NEGATIVE NUMBERS
Page 26
POWERS, ROOTS & LOGARITHMS
Page 34
MODULAR ARITHMETIC
Page 42

FOCUS | *RSA was the first significant public key/private key system, and is the best known. Its name references Ron Rivest, Adi Shamir, and Leonard Adleman, the American and Israeli mathematicians who developed it in 1978. It wasn't realized at the time, but the RSA system had already been invented in 1973 by British mathematician Clifford Cocks. Cocks had been working at GCHQ, the British intelligence encryption facility, so his idea had not been publicized.*

PROBABILITY

THE MAIN CONCEPT | Applying mathematics to games of chance provided the first opportunity to consider probability, a mathematical analysis of chance, which initially gave it a bad reputation. It doesn't help that probabilities can seem unnatural. For example, if a coin comes up heads ten times in a row, it's almost impossible not to feel that the next toss is more likely to be tails. In fact, the coin has no memory; there is a 50:50 chance of heads or tails, whatever was thrown before. In the sixteenth century, Girolamo Cardano extended the simple probability of a single event to take in multiple occurrences. Probability is usually stated as a value between 0 (no chance) and 1 (will definitely happen). For example, if there's a 1 in 6 (or $\frac{1}{6}$) chance of throwing a 5, with a die, the chance of throwing two 5s in a row is $\frac{1}{6} \times \frac{1}{6} = \frac{1}{36}$. The chance of throwing a 4 or a 5 on a single throw is $\frac{1}{6} + \frac{1}{6} = \frac{2}{6}$, or $\frac{1}{3}$. Probability theory was expanded to take in distributions of probability and more sophisticated aspects of random events, and would be crucial to the development of quantum physics, where the behavior of quantum particles is directly controlled by probability.

DRILL DOWN | Cardano's greatest discovery in probability was the chance of throwing, say, a 6 with *either* of two dice. Naively, this could be $\frac{1}{6} + \frac{1}{6}$. But that would mean with six dice you would have $6 \times \frac{1}{6} = 1$; in other words, you would definitely get a 6 if you threw six dice. Cardano realized that he knew the probability of *not* throwing a six with *both* of two dice. The chance of not throwing a six with one die is $\frac{5}{6}$, so the chance of doing this twice is $\frac{5}{6} \times \frac{5}{6} = \frac{25}{36}$. What was left—$1 - \frac{25}{36}$ (or $\frac{11}{36}$)—was the chance of getting a 6 with either die.

FOCUS | *Probability can even mislead mathematicians. When in 1990 Marilyn vos Savant, writing for* Parade *magazine, posed the so-called Monty Hall problem (see page 139), she was deluged with mail from mathematicians and scientists telling her she was wrong. She wasn't. Perhaps the best letter said: "You're wrong, but look on the positive side. If all those PhDs were wrong, the country would be in very serious trouble."*

STATISTICS

THE MAIN CONCEPT | The term "statistics" used to apply to a collection of data about a state—a country—the kind of thing you might find in the CIA World Factbook. However, by the seventeenth century, thanks to the work of an English button maker called John Graunt, statistics took on a new form. Graunt used data from "Bills of Mortality" in London, plus any information he could gather about births, to help understand the ebb and flow of the city's population and the spread of disease. Statistics takes data from a collection of individuals and produces collective information, often extrapolating from what is known to make predictions. One of Graunt's early efforts was to try to work out life expectancies for different groups of people, which would become the foundation of the insurance business. Statistics is not limited to human subjects, but has proved a powerful tool for assessing the behavior of any large mass of similar objects. In the nineteenth century, "statistical mechanics" was used to understand the behavior of collections of gas molecules, explaining phenomena such as gas pressure and viscosity. By this time, basic collection and collation of statistics (as, for example, in a census) had been combined with probability (see page 124) to make predictions on the future behavior of everything from stocks and shares to voters in an election.

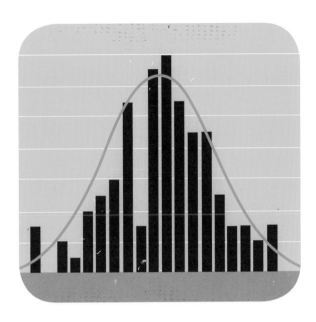

DRILL DOWN | The basic collection of statistics is not controversial, but when it is used to make predictions, a statistical approach can be misleading. Weather forecasting is a good example, where the system is so complex that it behaves in a way that mathematicians describe as "chaotic," making any prediction prone to error. Similarly, the forecasts that businesses use in their budgets often result in long postmortems as to why a company has not behaved as forecast— in practice, the real question should be the accuracy of the forecast. Statistical forecasting is very valuable, but we rarely have enough information on potential errors to make most effective use of the data.

FOCUS | *There is some confusion over who first produced the most famous quote on statistics (and one of the most famous in all of mathematics): "There are three kinds of lies: lies, damned lies, and statistics." It is often attributed to British prime minister Benjamin Disraeli. He claimed to be quoting Mark Twain, but the remark does not appear in Twain's writing.*

FACTORIALS & PERMUTATIONS
Page 96
PROBABILITY
Page 124
OPERATIONS RESEARCH
Page 140

LOGARITHMS & THE SLIDE RULE

THE MAIN CONCEPT | We are so used to computers that it is difficult to remember the long period in history when calculations that did not need to be exactly worked out by hand were undertaken using logarithms or with the mechanical device for logarithm manipulation, the slide rule. The slide rule, sometimes called a "slipstick," was an analog computer—one that used physical values rather than digital ones. Basic slide rules could multiply and divide, while more sophisticated devices had features such as roots, sines, and cosines. The standard slide-rule design had a pair of rulerlike sections, which were fixed parallel to each other, with a third sliding section positioned between them. The sliding section was moved until a pointer lined up with one value for the calculation on the top section, then the answer would be read off from one of the fixed sections. The final part of the slide rule was the cursor, a transparent part that was slid along above the three sections. This had a line running top to bottom that made it possible to read off values on nonadjacent scales. Because the scales were logarithmic—instead of being marked off regularly as on a ruler, values came closer and closer together—they produced multiplication or division by what was effectively addition or subtraction.

DRILL DOWN | A simple slide rule operation might be to multiply 2.5 by 2.8. The center rule would be slid along until the 1 marker on one of its scales lined up with 2.5 on the scale of the rule above it. The user would then move the cursor until it lined up with 2.8 on the center rule's scale and would read off the answer from the upper rule's scale, which would be positioned over 7. This worked because the center rule was moved along by a distance of log 2.5 and then a distance of log 2.8 was added on. Because with a logarithm $log(n) + log(m) = log(nm)$, the value read off was the correct answer.

NUMBER BASES
Page 24
POWERS, ROOTS & LOGARITHMS
Page 34
CALCULATING MACHINES
Page 130

FOCUS | *The first slide rules were developed in the 1620s by William Oughtred, an English mathematician. He took two existing rulers with logarithmic markings and slid them along each other to perform calculations. He then published a design for a circular slide rule (an approach that was never as popular as straight rules), but the modern style would not be introduced until the 1850s.*

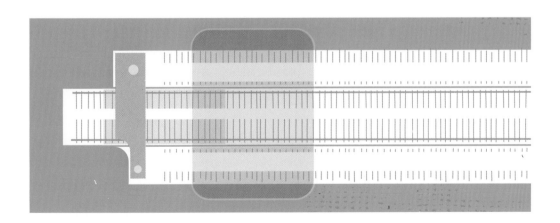

CALCULATING MACHINES

THE MAIN CONCEPT | Calculations using stones as markers were common in ancient times (hence, for example, "calculus," which means "small stone" in Latin). But when those stones were transformed into beads on wires or rods, they became what is arguably the prototypical calculating machine, the abacus. In effect, an abacus simulates the columns of position-based arithmetic, allowing a user to rapidly add and subtract, with techniques available to enable multiplication and division. But the idea of using gearing to undertake mathematical functions seems to have emerged alongside the improvement of clocks in the seventeenth century. There is some dispute over who got there first, but French mathematician Blaise Pascal produced a mechanical calculator in 1642 when he was just 19—dozens were made, but they were clumsy and prone to error. The first mechanical calculator in large-scale use was French inventor Charles Thomas's arithmometer, which became a standard from the 1850s: it was reliable and capable of true multiplication rather than repeated addition. This came after English mathematician Charles Babbage's incomplete designs for a more sophisticated calculator (the Difference Engine) and programmable computer (Analytical Engine). Mechanical calculators remained in use until replaced by electronic calculators in the 1970s.

DRILL DOWN | The Difference Engine, a small part of which was demonstrated by Babbage in 1822, was designed to work on up to seven numbers at a time and was intended to help with the automation of production of mathematical, astronomical, and navigational tables, which required tedious repeated manual calculation. Babbage lost interest in it when he put together his concept of the Analytical Engine in 1837, which would have been a programmable mechanical computer incorporating the components of modern computers in mechanical form, with data input on punched cards originally designed to control mechanical looms. Mathematician Ada Lovelace devised potential programs for the Analytical Engine in her notes on it.

FOCUS | *The oldest known calculating machine is the Antikythera Mechanism. Discovered in an ancient Greek shipwreck off the island of Antikythera in 1901, the mechanism was not properly investigated until the 1970s, when it was examined with X-rays and gamma rays and shown to be an analog mechanical astronomical calculator with at least 30 gears, and around 2,200 years old.*

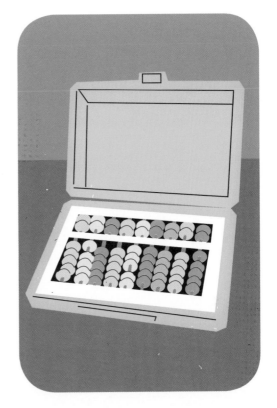

FOURIER TRANSFORMS

THE MAIN CONCEPT | Few techniques in applied mathematics are as impressive in their power as Fourier transforms. Named after French mathematician Joseph Fourier (1768–1830), the Fourier transform is, in essence, a powerful technique for breaking a complex mathematical structure down into a collection of simple ones. Strictly speaking, there are three Fourier entities. The Fourier series is a potentially infinite set of simple sine waves that can be combined to make up a more complex mathematical function, provided that function is continuous, rather than having sudden, discontinuous leaps. For example, in the diagram opposite, sine waves are being combined to come increasingly closer to the very different-looking square wave. The various kinds of Fourier transform are mechanisms for decomposing a function into its components, and Fourier analysis is the broader application of these kinds of techniques. One widely used variant, the fast Fourier transform, is particularly flexible as it works on a sample of the data. Although the most obvious application of Fourier transforms is in sound—for example, building up different sounds in a synthesizer by combining simple sine waves—Fourier analysis has found use everywhere from physics to the software for digital cameras to stock markets and the analysis of the structure of proteins.

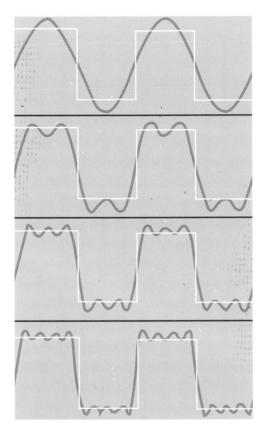

DRILL DOWN | When German mathematician Carl Gauss was working on the orbits of asteroids in the early 1800s, using relatively small samples of data, he developed a technique that would only be rediscovered in full in the 1960s. The fast Fourier transform is, as the name suggests, a fast algorithm for calculating a discrete Fourier transform—one that works on a limited set of samples of data, each taken at the same interval, dividing each sample into its separate frequency components. Most of us regularly use the output of fast Fourier transforms, as they are involved in the compression algorithms in JPEG images and MP3 sound files.

FOCUS | *Physicist Max Tegmark and astrophysicist Matias Zaldarriaga have suggested it would be possible to be build a radio telescope using an array of detectors, each performing fast Fourier transforms on the data, which would give it the flexibility of a single dish telescope but the ability to cover a wide field of view, like the much more expensive interferometer observatories featuring multiple telescopes.*

MATRIX MANIPULATION

THE MAIN CONCEPT | Matrices (or matrixes—both plurals of "matrix" are used) are two-dimensional rectangular arrays. They can be thought of as a mathematical chest of drawers that can either have several drawers in a single row, several drawers in a single column, or both rows and columns of drawers. Each row must have the same number of drawers. Similarly, columns are each the same size. At its simplest, each drawer can contain a number, but it can also hold functions. The power of a matrix is that it allows mathematical operations to be applied to a whole collection of values all at once. Matrix addition requires the two matrices to be the same shape, adding each element of one to the equivalent element of the other. Matrix multiplication is more interesting. It requires the first matrix to have the same number of columns as the second matrix has rows. This is because each value in the resultant matrix is calculated by adding together the result of multiplying the elements in the equivalent row of the first matrix by the elements in the equivalent column of the second matrix. If both matrixes are the same shape, it's possible to get two different results depending on the order of multiplication.

DRILL DOWN | In simple arithmetic, multiplication is commutative. This means that $A \times B = B \times A$. This seems perfectly natural—so much so that it's hard to see that this isn't always going to be the case. Even in basic arithmetic, though, other operations, such as subtraction, are not commutative. $A - B \neq B - A$. In general, multiplication also need not be commutative—as is the case for matrices. Only square matrices can be multiplied in both directions, because of the need for rows and columns to be equivalent. But unless the matrices are suitably symmetric, reversing the order in which the matrices are multiplied produces different answers.

FOCUS | *One of the strangest computer-programming languages ever, APL (standing for "A Programming Language") was developed in the late 1950s and early 1960s. Most programming languages consist of a series of English-like statements, but APL uses matrix manipulation to handle whole matrices at a time, producing very compact, mathematics-like code. The entire program to generate all the prime numbers up to N, for example, would be (~N∈N∘.×N)/N←1↓ιN.*

GAME THEORY

THE MAIN CONCEPT | Game theory sounds like the rules of football, but uses simple decision-based games to explore human behavior. The games often involve two players, with opportunities to cooperate or to try to get one over on the other player, often with no knowledge of the other player's decision. Such games are frequently used by economists and psychologists. John von Neumann introduced the discipline in 1928 with his paper *On the Theory of Parlor Games*, while American mathematician John Nash—featured in the movie *A Beautiful Mind*—developed the field significantly. The most studied games are the Prisoner's Dilemma and the Ultimatum Game. The latter demonstrates how humans will sacrifice themselves to punish someone else. The Prisoner's Dilemma was the basis of "mutual assured destruction" in the Cold War. It imagines two prisoners, each of whom, without communicating, can either support the other or betray them. If only one betrays, he or she is let free and the other gets a long sentence. If both support, they each get a short sentence. And if both betray, each gets a mid-length sentence. The combined benefit is highest if both support, but an individual benefits by betraying, as long as the other doesn't. It's logical for an individual to betray—but if both follow that logic, they suffer.

DRILL DOWN | The widely used Ultimatum Game distributes a sum of money between two people. The first person decides how to split the money, while the second either accepts the split or refuses it, in which case neither gets anything. Players from Western cultures usually turn down offers less than about 35 percent. Economically this is illogical—it's refusing to take free money—but psychologically it puts a price on punishing the other player for being greedy. What is rarely observed is that with a large reward (the game is usually played for amounts in the $1–10 range), this percentage split no longer holds, as players will accept a much smaller percentage.

FOCUS | *A game theory term in wide use is "zero-sum game." This is a game in which the losses of some players are canceled out by the winnings of the other players: adding gains and losses gives zero. It rules out the possibility of win-win. If you are sharing candies, for instance, and give someone too many, someone else gets too few.*

MATHEMATICAL LOGIC
Page 44
GRAPH THEORY
Page 68
FACTORIALS & PERMUTATIONS
Page 96

MONTE CARLO METHODS

THE MAIN CONCEPT | Although probability has largely moved out of the shadow of gambling, it can't escape its origins entirely: fair games of chance are the simplest examples of probability in action. Nowhere is this more obvious than in "Monte Carlo methods," unashamedly named after the Mediterranean casino. However, despite the name, Monte Carlo methods, or simulations, are not about the best way to win at roulette, but rather a mechanism for mathematically simulating a small part of reality. The approach involves harnessing the power of randomness to make predictions. "Monte Carlo" was used as a codeword for the technique when it was developed by Stanislaw Ulam and John von Neumann, working on the Manhattan Project to develop nuclear weapons during World War II. The first application was to work out requirements for radiation shielding to stop neutrons. Because the interactions of particles is probabilistic, by repeatedly using a random selection of parameters that fit the behavior of the neutron, a picture of possible outcomes was built up. Monte Carlo simulations are often used in fields from finance to physics, in situations where the environment is too complex to develop an effective deterministic model (one that has definite values), but where it is possible to apply probabilities and statistical data to the variables.

DRILL DOWN | A simple Monte Carlo simulation has been used to prove the outcome in the "Monty Hall problem." This involves a game where a contestant chooses between three doors and wins what is behind the chosen door—two doors have goats behind them, while the other conceals a car. After the contestant makes a choice, the host opens one of the other doors to reveal a goat. The question is, should the contestant stick with their choice or shift to the other unopened door? The counterintuitive reality is that they are twice as likely to win if they switch. This seems so unlikely it was often simulated by repeatedly running the game for the "stick" and "switch" options, proving the result.

FOCUS | *Monte Carlo methods need random numbers, which aren't easy to generate. True random values can be produced using quantum devices, but typical computer "random" values use a pseudo-random series. A simple example takes the previous value (the first value is "seeded" using, say, the time), multiplies it by a large constant, adds another large constant, and takes the modulus to the base of a third large number.*

OPERATIONS RESEARCH

THE MAIN CONCEPT | Also known as operational research, this is a discipline that was developed in the lead-up to World War II to apply mathematical methods to solving military operational problems, from the most effective deployment of depth charges to the best routes to take on a multi-leg journey. As well as using standard statistical techniques and Monte Carlo methods (see page 138), operations research incorporates methods such as queuing theory, which is used to minimize the time customers (or things) spend in a queue; linear programming, which maximizes a value, such as the profit of a transaction, given a range of constraints; and dynamic programming, which breaks down a problem into smaller components that can be dealt with using recursion (see opposite). Operations research is now also used in business, to deal with large, complex problems that don't easily fit simple mathematical solutions. Operations research was originally a purely mathematical discipline, but it was an early adopter of computers and now it is highly dependent on computing software, from spreadsheets to intricate visual simulations. The term "optimization," implying finding the best possible outcome, is often associated with operations research, although in practice because of the messy nature of the problems they are applied to, many techniques produce near-optimization.

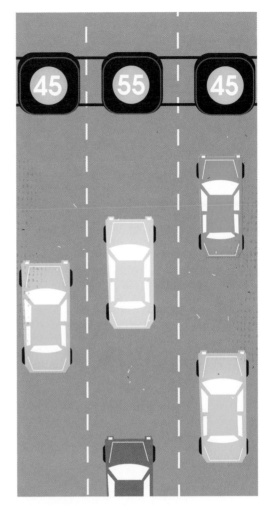

DRILL DOWN | Recursion frequently occurs in mathematics and computing and is a powerful tool in some of the algorithms that are employed in operations research. Recursion requires a starting and stopping point along with a rule that allows the next instance to be generated. Crucially, recursive algorithms are defined in terms of themselves, which enables them to be compact for the work they do. For example, the factorial of n is "n multiplied by the factorial of $n-1$." Which in turn is "$n-1$ multiplied by the factorial of $n-1-1$," etc. Recursion enables a simple rule to build up a significant outcome, and a set of recursive rules can produce complex behavior.

FOCUS | *Operations research can help make the best choice in hiring someone (or marrying someone). This kind of problem, known as an "optimal stopping problem," involves choosing something from a sequence without being able to see the whole sequence. With a total of* n *candidates, you should reject the first* √n *people and choose the next person better than any of those.*

PROBABILITY
Page 124
STATISTICS
Page 126
MONTE CARLO METHODS
Page 138

COMPUTERS

THE MAIN CONCEPT | The electronic programmable computer transformed both the ability to perform arithmetic and much more sophisticated mathematical calculations. In effect, computers divided mathematics into two, with some mathematicians still using paper and pencil, and others adopting computer methods. The early electronic computers were based on thermionic valves, also known as vacuum tubes. The first programmable computer is now accepted to be the Colossus, built at Bletchley Park in the UK as part of the effort to decipher German military communications. Colossus went live at the start of 1944, but its role was not originally recognized, as the Colossus devices were destroyed after the war due to an unfortunate idea of military security. The more flexible ENIAC computer in the US went into use a year after Colossus and became the model for early developments. Such early computers were room-sized and used as much electricity as a modern office building. Programming was initially done with switches on the device, then punched paper tapes and cards, before moving to teletypes and display screens. Although such "mainframe" computers became much more powerful and relatively compact for that power, they remained corporate in scale until the introduction of personal computers in the 1980s. Now, a smartphone has far more computing power than the early mainframes.

DRILL DOWN | The valves (now transistors) in electronic computers are arranged in logic circuits called "gates." Each gate provides a single operator from Boolean algebra, such as NOT, OR, and AND. In the gate, the principle electronic components act as switches that control one signal with another. If the signal is active it is treated as 1, or 0 if it is not active. So, for instance, a NOT gate, which switches 0 to 1 and 1 to 0, will output a signal if there is no input signal, but won't output anything if there is an input signal. All the other gates apart from NOT have two inputs controlling a single output.

FOCUS | *It is sometimes claimed that the first bug in a computer was in one of the electromechanical predecessors of electronic computers, recorded by US computer scientist Grace Hopper, who in 1947 attached a moth found in the computer to the computer's log book. However, the term "bug" for a technical problem had been used in engineering since the 1870s.*

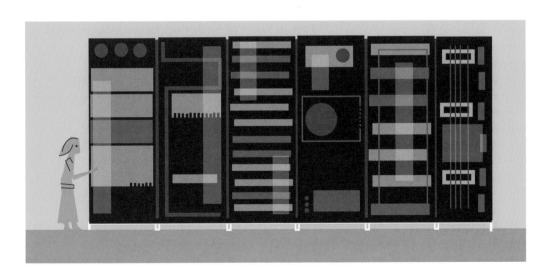

CHAOS THEORY

THE MAIN CONCEPT | Chaos sounds like a totally random concept, but mathematically speaking, a chaotic system is one that obeys clear rules—there's nothing random—but the system is either sufficiently complex or strongly interacting that very small changes can powerfully influence the outcome, meaning that it can be very difficult to predict what is going to happen next. Chaotic systems can be surprisingly simple. A pendulum consisting of two rods, one hinged on the bottom of the other, behaves impressively chaotically. Perhaps the best-known chaotic system is the weather—here it is the interaction of a whole range of weather systems that produces chaos. It was when mathematician Edward Lorenz was attempting an early computerized weather forecast that he made the first steps in chaos theory. He had printed off the numbers he needed to input into his computer program partway through the run—it took a long time and he wanted to be able to restart it at this point. But when he did so, the forecast was totally different. He realized that the printout used fewer decimal places than the program; just a tiny fraction of a difference in the values transformed the forecast. In addition to the weather, chaos turns up in everything from stock market behavior to engineering.

DRILL DOWN | Chaos and randomness each produce unpredictable behavior, but in different ways. A chaotic system is in theory entirely predictable, but is so sensitive to slight changes in the setup that in practice it behaves in a wild fashion. This gets amplified with time so, for example, weather forecasts are impossible to extend for more than about seven days. Random behavior is inherently unpredictable at the level of individual changes, but can often be predicted statistically, as many random actions have known probabilities. Some chaotic systems interact in such a way that they naturally tend toward certain possible values. When this occurs, the values the system tends toward are known as "strange attractors."

FOCUS | *The originator of chaos theory, Edward Lorenz was also responsible for the most dramatic image attached to the theory when he wrote the paper "Does the flap of a butterfly's wings in Brazil set off a tornado in Texas?" This "butterfly effect" became widely known, even though Lorenz's conclusion was, "No, it doesn't."*

COMPLEXITY THEORY

THE MAIN CONCEPT | In general use, "complexity" simply means having a structure with lots of parts, or that is particularly intricate. But in mathematics and computation, complexity describes a system with multiple interacting subsystems that results in "emergent" properties: the whole is greater than the sum of the parts. A living organism is complex in this sense. It is made up of individual cells, none of which are capable of the abilities of the whole. Often the behavior of a complex system will be nonlinear—having sudden, unexpected changes—and can be chaotic (see page 144). Complexity is probably most studied in computational mathematics, but it is also used in fields such as economics, biology, and network theory. One complexity mechanism is the "feedback loop." This is where a property of a system can make direct changes to the behavior of that system. Feedback can either run away (positive feedback) or dampen down behavior (negative feedback). The most familiar positive feedback is the squeal of a sound system when a microphone is too close to a loudspeaker, so that background noise is picked up, amplified, picked up even louder, and so on. Negative feedback was first studied in the mechanism of governors for steam engines. These mechanical devices are driven by the pressure in the engine. If they rotate too fast, they release steam, automatically reducing the pressure.

DRILL DOWN | In computing, complexity describes the difficulty of solving a problem. One that can be solved in polynomial time (when the maximum time taken is proportional to a power of the number of components in the problem—say, n^2 time units, where n things are involved) is called class "P." By contrast, an "NP" problem's solution can only be verified in polynomial time. It is thought (though not proved) that solving a third class of problem, "NP-hard," in polynomial time is impossible. Such a problem takes exponential time—say, 2^n time units—to solve. Usually, such problems only have approximate solutions. A typical NP-hard problem is the traveling salesman problem, finding the best route between a number of cities.

FOCUS | *Spontaneous order, or "self-organization," is a characteristic of complexity. A layer of wax on a board can demonstrate such complexity. Pour hot water down the angled sheet of wax and initially rivulets will run all over it. But as the water melts channels, more water will run down the channels, which will get wider still. The pattern formed is self-organizing.*

GLOSSARY

ALGORITHM—a series of instructions and rules used to carry out a task in a systematic fashion. Commonly applied to the logical structure of a computer program.

ARABIC NUMERAL—the worldwide standard for representing numbers (0123456789). Of Indian origin, but often called "Arabic" as they came to Europe via the Middle East.

AXIOM—a mathematical assumption that is so obvious that it need not be proved.

BASE—the number of values that can appear in a column when writing out a number. We usually work to base 10, with values 0 to 9 in the first column before starting a new one.

BINARY—numbers written to base 2, used by computers. The number 37 is binary 100101.

BOOLEAN [ALGEBRA]—a symbolic approach to logic. It combines values of "true" or "false" using the terms AND, OR, and NOT.

CALCULUS—the mathematics of change. Calculus comes in two forms—differential calculus (which describes how one variable changes with another) and integral calculus (which combines changing values to, for example, work out areas of shapes). The two forms are inverses of each other.

CARDINALITY—the size of a set. Two sets have the same cardinality if each member of one set can be paired off with a member of the other set with no members of either left over.

CHECKSUM—a digit used to check that values in a number have been typed correctly. The last digit of a credit card number is calculated from the other numbers, providing a checksum.

COMPOSITE NUMBER—a positive integer that is made by multiplying two smaller positive integers. It can be divided by one, itself, and at least one other positive integer.

CONGRUENT—geometrical shapes that have the same size and shape.

CONSTANT—a value in an equation or expression that is a fixed number. E.g., in the equation $3x + 5 = 0$, the value 5 is a constant.

COSINE—the cosine of an angle is the ratio in a right-angled triangle of the length of the side adjacent to the angle (other than the hypotenuse) to the hypotenuse.

ELLIPSE—the shape produced by drawing around two focal points such that the sum of the distances of a point on the circumference from the two focal points is constant, or one of the shapes produced by slicing through a cone. A circle is a special case of an ellipse where the focal points overlap.

ENCRYPTION—the mechanism for concealing values (numbers or letters) by a mathematical process to keep those values secret. The reverse process, uncovering the values, is "decryption."

EQUATION—a mathematical structure with two parts, where the value of one part is equal to the other. Each part may contain any combination of mathematical terms.

EXPONENT—when a variable or number is raised to a power, the number representing that power is the exponent. For example, in the equation $E=mc^2$, the number 2 is the exponent of c.

EXPRESSION—a combination of constants, variables, functions, and so on, with symbols showing their relationships.

FACTOR—any one of a set of numbers that are multiplied together to make another number.

FACTORIAL—a sequence of decreasing sized integers multiplied together, represented by ! For example, 5! is $5 \times 4 \times 3 \times 2 \times 1$.

FIELD—a property that has a defined value for each point in space and time.

FRACTION—a part of an integer. Rational fractions can be represented as a ratio of integers, such as $1/2$ and $37/159$, or decimals such as 0.5 and 0.2327044 . . . Irrational fractions can only be represented as decimals.

FUNCTION—a shorthand way to write out a mathematical calculation that can be applied to any value. So $f(x)$ applies the function f—anything from "multiply by 2" to a complex formula—to whatever x is.

GATE—a logical building block of a computer, which switches electrical currents according to a Boolean operator such as AND, OR, or NOT.

GOLDEN RATIO—a ratio where the ratio of the larger number to the smaller value is the same as the ratio of the sum of the two numbers to the larger value. This appears in nature and has been used by many artists, who believe it produces visually pleasing arrangements.

GROUP—a set with a mechanism for combining any two elements of the set to produce a third. For example, the integers form a group with addition as the mechanism.

HYPERBOLA—a mathematical curve like a pair of letter U shapes (with the two arms heading away from each other on each U) and with their points facing each other. One of the shapes produced by slicing through a pair of cones, placed point to point.

HYPOTENUSE—the side of a right triangle that is opposite the right angle.

INTEGER—a whole number such as −3, 1, or 55.

IRRATIONAL NUMBER—a number that cannot be made up of a ratio of two whole numbers. The square root of 2 is irrational.

LOGIC GATE—*see* gate.

MAINFRAME—a large central computer, usually requiring special environmental conditions and often shared between many simultaneous users.

MANIFOLD—a multidimensional geometric structure locally analogous to traditional "flat" Euclidian geometry. The surface of a 3D object such as a sphere, for example, is a 2D manifold

because any small part of it resembles a flat plane.

MATRIX—a rectangular collection of numbers in columns and rows.

MODULUS—when we use a system of numbers that reaches a value and goes back to the start, the modulus is the maximum value. For example, on a 12-hour clock the modulus is 12—after counting up to 12, the next value is 1.

NUMBER LINE—an imaginary line that is like an infinite ruler running from −∞ at one end to ∞ at the other with zero in the middle.

OPERATOR—a mechanism that is used to apply a function or change to one or more variables simultaneously. Simple operators include the Boolean operators AND, OR, and NOT, but operators can also apply complex formulae.

PARABOLA—a mathematical curve shaped like a U but with the arms heading away from each other. One of the shapes produced by slicing through a cone.

PI—a mathematical constant, pi (π) is the circumference of a circle divided by its diameter. The number begins 3.14159 . . .

POLYNOMIAL—an expression that only contains constants and variables, which can only be added, subtracted, or multiplied.

PRIME NUMBER—a positive integer that is bigger than 1, divisible only by 1 and itself.

PROOF—a collection of logical steps starting from axioms and making a step-by-step progression to a conclusion.

QUADRATIC EQUATION—an equation in the form $ax^2 + bx + c = 0$.

RADIAN—a unit used for measuring an angle or amount of rotation; an alternative to degrees—360 degrees is 2π radians.

RATIO—a ratio describes the relative size of two numbers. For example, 3:1, a ratio of 3 to 1, indicates that the first number is three times bigger than the second. Similar to a rational fraction, where $a{:}b$ is equivalent to a/b.

SCALAR—a value described by a simple number. For example, speed is a scalar.

SET—a collection of numbers, objects, or concepts. Sets provide the foundation of arithmetic, and much of mathematics can be built from them.

SIMULTANEOUS EQUATIONS—two or more equations that can be combined to provide information about the behavior of multiple variables.

SINE—the sine of an angle is the ratio in a right triangle of the length of the side opposite the angle to the hypotenuse.

SQUARE ROOT—the number that, multiplied by itself, produces the value of which it is a square root. The square root of 9 is 3, as $3 \times 3 = 9$.

TANGENT—a straight line drawn just touching a curve, and which has the same slope as the curve at that point.

THEOREM—a statement that has been proven with a formal mathematical proof.

TRANSCENDENTAL—a number, such as π, that cannot be calculated by a finite formula.

VARIABLE—a value in an equation or expression that can be changed to any value. For example, in the equation $3x + 5 = 0$, the value x is a variable.

VECTOR—a value that has both size and direction. For example, velocity is a vector as it combines a speed with the direction of that speed.

VENN DIAGRAM—a diagram devised by mathematician John Venn to show the logical relationships between different sets.

VERTEX—a point where the edges of a shape meet.

FURTHER READING

Books

Acheson, David. *1089 and All That.* Oxford: Oxford University Press, 2010.

A refreshing exploration of the joy of mathematics, from chaos theory to the Indian rope trick.

Aczel, Amir. *Finding Zero.* New York: St. Martin's Press, 2015.

A personal odyssey to discover the origins of zero.

Bellos, Alex. *Alex's Adventures in Numberland.* London: Bloomsbury, 2011.

A confection of mathematical experiences, stretching from casinos to the world's fastest mental calculators.

Blastland, Michael, and Andrew Dilnot. *The Tiger that Isn't.* London: Profile Books, 2007.

A wonderful exploration of how statistics and numbers in general have been used to mislead.

Cheng, Eugenia. *Cakes, Custard and Category Theory.* London: Profile Books, 2015.

A journey through the mind of the mathematician, incorporating Cheng's specialty, category theory.

Christian, Brian, and Tom Griffiths. *Algorithms to Live By.* William Collins, 2016.

Takes the concept of algorithms and shows how the math can be used in real life.

Clegg, Brian. *Are Numbers Real?* New York: St. Martin's Press, 2016.

A history of mathematics, showing how it has gradually become more detached from reality.

Clegg, Brian. *A Brief History of Infinity.* London: Constable and Robinson, 2003.

A history of the most mind-boggling aspect of mathematics, through the people involved in developing the concept.

Clegg, Brian. *Dice World.* London: Icon Books, 2013.

The influence of randomness and probability on our world and lives.

Clegg, Brian, and Oliver Pugh. *Introducing Infinity.* London: Icon Books, 2012.

An entertaining graphic guide to the concept of infinity.

Du Sautoy, Marcus. *The Number Mysteries*. London: Fourth Estate, 2011.

Balances five of the great unsolved mathematical mysteries with the practical applications of math in real life.

Gardner, Martin. *Mathematical Puzzles and Diversions*. Chicago: University of Chicago Press, 1961.

The classic book of recreational mathematics and its follow-up titles—including the latest, *My Best Mathematical and Logic Puzzles* (2016)—are still superbly entertaining.

Gessen, Masha. *Perfect Rigour: A Genius and the Mathematical Breathrough of the Century*. London: Icon Books, 2011.

The story of Russian mathematician Grigori Perelman, who solved one of the great mathematical challenges, the Poincaré conjecture, only to drop out of mathematics altogether and turn down a $1 million prize.

Gleick, James. *Chaos: Making a New Science*. London: Vintage, 1988.

A very journalistic and readable book on the development of chaos theory.

Hayes, Brian. *Foolproof*. Cambridge, Massachusetts: MIT Press, 2017.

A range of articles on fascinating math topics, from random walks to the story of the mathematician Gauss's feat of adding 100 numbers instantly.

Hofstadter, Douglas. *Gödel, Escher, Bach*. New York: Basic Books, 1979.

Classic, and to many mystifying, book on the essence of mathematics and cognition, but if it works for you, superb.

Livio, Mario. *The Equation that Couldn't Be Solved*. New York: Simon and Schuster, 2005.

A surprisingly engaging history of algebra and the development of group theory.

MacCormick, John. *Nine Algorithms that Changed the Future*. Princeton: Princeton University Press, 2012.

An exploration of some of the key algorithms that shape our online world, from Google's PageRank to the cryptography that keeps data safe.

Mackenzie, Dana. *The Story of Mathematics in 24 Equations*. London: Modern Books, 2018.

Uses 24 important equations through history to show how mathematics has developed.

Nicholson, Matt. *When Computing Got Personal*. Bristol, UK: Matt Publishing, 2014.

An excellent history of personal computing.

Parker, Matt. *Things to Do and Make in the Fourth Dimension*. London: Penguin, 2015.

Stand-up mathematician Parker presents a range of recreational math, from interesting ways to divide pizza to the importance of the 196,883rd dimension.

Petzold, Charles. *Code*. Redmond, Washington: Microsoft Press, 2000.

Using familiar aspects of language, this Windows expert uncovers the inner workings of computer programs for the general reader.

Scheinerman, Edward. *The Mathematics Lover's Companion*. Newhaven, Connecticut: Yale University Press, 2017.

Takes on 23 of the more interesting subjects of mathematics, from prime numbers to infinity, in some depth but still readable.

Singh, Simon. *Fermat's Last Theorem*. London: Fourth Estate, 1997.

The remarkable story of how a mathematical puzzle posed in the seventeenth century dominated the life of a twentieth-century mathematician.

Stewart, Ian. *The Great Mathematical Problems*. London: Profile Books, 2013.

A collection of some of the greatest challenges to face mathematicians through history.

Stewart, Ian. *Professor Stewart's Cabinet of Mathematical Curiosities*. London: Profile Books, 2010.

The best of mathematician Stewart's compendia of puzzles, odd mathematical facts, and more.

Stewart, Ian. *Significant Figures*. London: Profile Books, 2017.

The lives of many great mathematicians.

Stipp, David. *A Most Elegant Equation*. New York: Basic Books, 2018.

Introduces Euler's remarkable equation $ei\pi + 1 = 0$ and explains why each of its main components is so important.

Watson, Ian. *The Universal Machine*. New York: Copernicus Books, 2012.

Pulls together the whole history of computing in an approachable form.

Websites

MacTutor History of Mathematics

Old-fashioned style but a huge number of biographies and history of mathematics topics

www-history.mcs.st-and.ac.uk

Mathigon

Well-designed, engaging tutorials in mathematics

mathigon.org

The Prime Pages

Everything and anything on prime numbers

primes.utm.edu

Quanta Magazine: Mathematics

Wide-ranging articles on the latest developments in mathematics

quantamagazine.org/tag/mathematics/

The Top 10 Martin Gardner Scientific American Articles

Great math writing from the king of recreational mathematics

blogs.scientificamerican.com/guest-blog/the-top-10-martin-gardner-scientific-american-articles/

Wolfram MathWorld

Wide-ranging mathematical resources from the leading mathematical software company

mathworld.wolfram.com

INDEX

ABOUT THE AUTHORS

Brian Clegg
With MAs in Natural Sciences from Cambridge University and
Operational Research from Lancaster University in the UK, Brian Clegg
(www.brianclegg.net) worked at British Airways for 17 years before
setting up his own creativity-training company. He now is a full-time
science writer with over 30 titles published, from *A Brief History of Infinity*
to *The Quantum Age*, and writes for publications from the *Wall Street
Journal* to *BBC Focus* magazine. He lives in Wiltshire, England.

Dr. Peet Morris
Peet Morris is a lecturer and researcher at the University of Oxford,
and also an Oxford alumnus (Keble and Wolfson Colleges). He works
in various fields, including Computational Linguistics (artificial
intelligence), Software Engineering, Statistics, and also
Experimental Psychology.

ACKNOWLEDGMENTS

Brian: For Gillian, Rebecca, and Chelsea.
Peet: For Harriet, and also my mathematical nephew, Ben.

With thanks to Tom Kitch and Angela Koo for setting this interesting challenge and helping bring it into being. A special thanks to Martin Gardner for his *Mathematical Puzzles and Diversions*, which showed that math could be fun.

Picture credits

The publisher would like to thank the following for permission to reproduce copyright material:

Alamy: photographer unknown 19 (right); Granger, NYC 121 (left)

LANL: 119 (top left), 121 (right). Unless otherwise indicated, this information has been authored by an employee or employees of the Los Alamos National Security, LLC (LANS), operator of the Los Alamos National Laboratory under Contract No. DE-AC52-06NA25396 with the U.S. Department of Energy. The U.S. Government has rights to use, reproduce, and distribute this information. The public may copy and use this information without charge, provided that this Notice and any statement of authorship are reproduced on all copies. Neither the Government nor LANS makes any warranty, express or implied, or assumes any liability or responsibility for the use of this information.

Shutterstock: Ron Dale 16 (top left); Tupungato 16 (top right), 18 (left); Peter Hermes Furian 16 (bottom right); Prachaya Roekdeethaweesab 19 (left); Georgios Kollidas 50 (bottom right); Nicku 51 (bottom left); Zita 51 (bottom right); Chiakto 84 (bottom left), 86 (left); Nicku 84 (bottom right), 86 (right); Rozilynn Mitchell 118 (top right); Bobb Klissourski 119 (bottom left)

Wellcome Collection (both CC BY 4.0): 17 (top right), 18 (right)

Wikimedia: Mark A. Wilson (Wilson44691, Department of Geology, the College of Wooster) (CC BY-SA 4.0) 52 (left)

All reasonable efforts have been made to trace copyright holders and to obtain their permission for the use of copyright material. The publisher apologizes for any errors or omissions in the list above and will gratefully incorporate any corrections in future reprints if notified.